Managing Variation for Injection Molding, 3rd Ed

I0474181

This book is written for green and black belt level personnel in the manufacturing industry. Most examples are directed toward injection molding of thermoplastic parts. This 3rd Edition includes several error corrections from original 2003 edition, more direction and discussion, and pages from the Basic Statistics and SPC book never published previously. Topics include:
- understand and quantify variation, understand sources of variation,
- detailed discussion of variable control charts,
- SQC, effective control charts, sub-grouping strategy, real time SPC,
- detailed discussion of attribute control charts,
- learn how to perform Gage R&R & MSE (measurement system evaluation),
- learn six sigma techniques, calculate Cpk, Ppk, understand Z-score math,
- calculate and perform correlation analysis,
- single & multi regression analysis to create predictive equations,
- use predictive regression equations to nominalize or improve dimensions,
- DOEs, ANOVA, COV (components of variance) - how to quantify % each,
- process mapping, process qualification & validation,
- nominalization,
- create scorecards to track performance, and more.

Copyright & ISBN

About the Author

The author - Jay Carender:

- Hands on processing skills,
- Mold build/mold design & engineering knowledge,
- Degree in Mechanical Engineering from GMI (General Motors Institute of Technology ... now known as Kettering University),
- DOE Training from Stat-Ease, Inc.,
- SPC Training from University of Tennessee Management Development Center,
- SPC Training from ASQC,
- Black Belt from AIT (Advanced Integrated Technologies),
- Productivity and Quality Improvement Training by Dr. Deming,
- Mold Design & Advanced Mold Design Training from New York University ... taught by John Klees Enterprises,
- Multiple courses on processing from RJG Industries, Inc.,
- Owner of Advanced Process Engineering - company started in 1990 to create and market pocket sized reference booklets for injection molding industry. Six booklets written & published along with other training manuals.

The hands on experience is from various fortune 500 companies performing injection molding. The extensive experience includes:

- Hot runner molding,
- High cavitation molds,
- High speed molding with cycle times less than 5 seconds,
- Stack molds, unscrewing molds, core pulls, slides, close tolerance parts,
- Engineering resins,
- SPC,
- Statistics,
- DOE to effect process improvement and dimensional nominalization.

Other APEBOOKS

Injection Molding Reference Guide, 4th Ed.
contains basic part design, trig tables, calculation for thermal expansion w/ coeffs, SHCS data, torque specs, shrink data, cooling equation, mold debug guidelines, melt index data, resin density data, many tables of process guidelines, process development techniques, calculating heat load & water flow requirements, pipe data, conversion factors, transformer & motor current, PM & safety, basic statistics, equip selection guidelines and more.

Injection Molding Troubleshooting Guide, 3rd Ed.
contains troubleshooting tips/solutions for many injection molding defects, intro to DOE, discussion of VPT and Decoupled Molding^{SM} techniques (SM - RJG, Inc). This 3rd ED. includes many select pages from other APEBOOKS which are applicable to process set up and troubleshooting such as sources of variation and root cause analysis.

Math Skills for Injection Molding, 2nd Ed.
contains intro to basic algebra, using conversion factors, percentages, ratios, proportions, trig as needed for draft and tapers, trig tables, thermal expansion calculations, calculate shrinkage, determining part cost, understanding efficiency and utilization, intensification ratios and clamp tonnage, projected area, residence time, cooling time, interpolation, heat load, Cp, Cpk, Pp, Ppk, correlation, math equations and samples for calculating piezo and strain gage transducer full scale pressure, and more.

Pocket Injection Mold Engineering Standards, 2nd Ed.
mold spec sheets, quoting & design direction, shrinkage, mold steels and hardness, heat treatment, thermal conductivity, thermal expansion, plating, surface finish tables, cooling design guidelines, gate designs, runner sizing, venting, sprue pullers, sucker pins, ejection, slides, support pillars, alignment guidelines, O-ring guidelines, hot runner info, torque specs, trig tables and more.

Managing Variation for Injection Molding, 3rd Ed.
now includes some basic statistics, understand & quantify variation, attribute and variable control charting discussion, techniques and formulas, 6 sigma techniques, Cpk, Ppk, Z-score math, correlation, single & multi regression analysis to create predictive equations, DOEs, ANOVA, components of variance - how to quantify % each, MSE & Gage R&R, SQC, real time SPC, process mapping, process qualification & validation, FMEA, nominalization, molding techniques to reduce variation, and more.

The Advanced Process Engineering Guide
a compilation of the first five APEBOOKS in one book!

For questions or comments,
contact Jay Carender and Advanced Process Engineering at: advproeng@gmail.com

Table of Contents

Table of Contents (continued)

Types of Variation

When there is variation, there exists conditions, components or parts which are unlike or dissimilar. These differences may be quite small. If the differences are smaller than the ability to discern such differences, we may incorrectly say the items in question are the same. The effect of such an incorrect classification of being the same may have no effect or a great adverse effect depending on the requirements for consistency.

There is always variation present to some degree, it just depends on our ability to recognize and quantify that variation. In the manufacturing world, we recognize variation by our ability to measure components. Such measurements are typically length, width, height, weight, strength, etc. The measurement data just described is characterized by a number value whether in inches, feet, pounds, millimeters, etc. Since these numbers can be compared to identify and quantify variation, such data is known as **variable data** (in contrast to **attribute data** which is go vs. no-go; good vs. bad).

In the manufacturing world, there are requirements which yield product fitness for use or basic functionality. These requirements are typically established by the customer (may be an internal or external customer). These requirements become known as the **voice of the customer**. The variation which exists during manufacture of the customer's parts becomes known as the **voice of the process**.

The variation described above as the voice of the process is typically normal variation as opposed to uniform variation seen in graphic below. Rolling just one die results in uniform variation (16.67% chance of getting any given number).

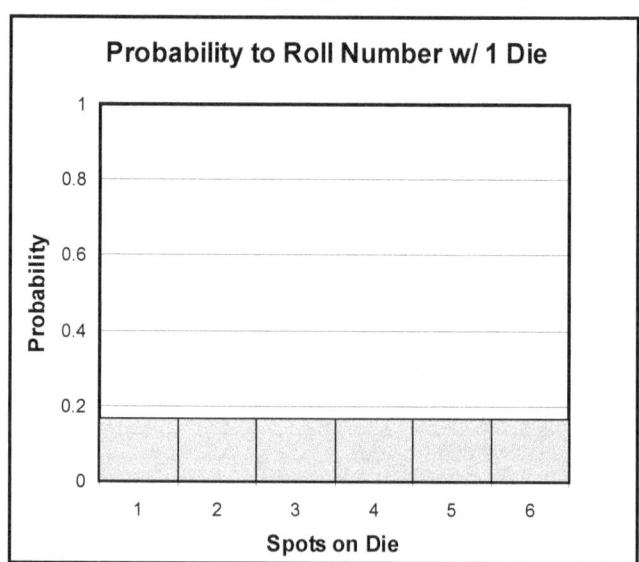

Fortunately, our manufacturing processes do not exhibit uniform variation ... if they did, we would have great difficulty achieving consistent and acceptable quality performance. Manufacturing processes do exhibit variation, but that variation is known as "normal" variation ... i.e. normal process variation which is characterized by a normal distribution curve (bell shaped curve).

Normal Distribution Curve

When you roll two dice, a table of combinations would look like the following:

	1	2	3	4	5	6
1	2	3	4	5	6	7
2	3	4	5	6	7	8
3	4	5	6	7	8	9
4	5	6	7	8	9	10
5	6	7	8	9	10	11
6	7	8	9	10	11	12

The resulting probability graph becomes pyramid shaped whereby chances for a seven are the greatest. In a manufacturing environment, the data will be normally distributed like the bell shaped curve superimposed on the two dice probability graph. This bell shaped curve is also called a normal distribution curve indicating <u>data to be clustered around a central mean</u>.

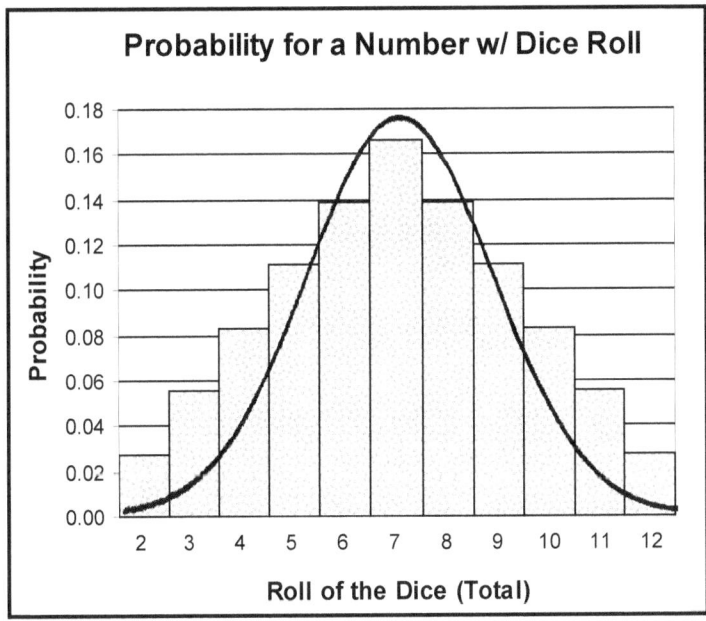

By applying the skills learned in this book, we can make this normal distribution more narrow; whereby, more of the data is clustered about the central target, and we can learn to define where our process mean is relative to the target. We can quantify this centering or nominalization with a single numeric term which helps to quantify performance. We can characterize the width of this normal distribution curve relative to specification limits with a single numeric term.

Variation w/ Real Data

The following groups of numbers include variation:

	Length (inches)
Hour 1	4.544
Hour 2	4.568
Hour 3	4.577
Hour 4	4.568
Hour 5	4.562
Hour 6	4.569
Hour 7	4.571
Hour 8	4.568
Hour 9	4.568
Hour 10	4.550
Hour 11	4.562
Hour 12	4.564
Hour 13	4.555
Hour 14	4.571
Hour 15	4.578
Hour 16	4.582
Hour 17	4.566
Hour 18	4.581
Hour 19	4.577
Hour 20	4.564
Hour 21	4.560
Hour 22	4.568
Hour 23	4.577
Hour 24	4.581

time		weight (grams)
7:00	1	962.46
7:00	2	964.12
7:00	3	963.57
10:00	1	964.66
10:00	2	965.11
10:00	3	963.14
13:00	1	966.75
13:00	2	965.14
13:00	3	965.18
16:00	1	965.44
16:00	2	964.98
16:00	3	965.66

	Drive Time (min)
Day 1	23.44
Day 2	22.98
Day 3	22.16
Day 4	21.43
Day 5	22.10
Day 6	22.52
Day 7	22.45
Day 8	22.82
Day 9	20.84
Day 10	22.04
Day 11	24.44
Day 12	23.66
Day 13	22.34
Day 14	24.16
Day 15	22.94
Day 16	22.15
Day 17	22.44
Day 18	21.06
Day 19	21.56
Day 20	24.36
Day 21	22.61
Day 22	21.69

Later in this book, we will learn how to calculate and quantify the above variation. With these techniques, we can know if one data set is better than another data set.

Before we quantify variation, we need to understand other terms which can cause variation or affect how we measure values. NOTE: In all data sets, some of the variation is real and some is caused by measurement error.

Cause and Effect Diagrams

What causes variation?

Any process, including driving to work, can be affected by many sources of variation. A cause & effect diagram is a good way to identify sources of variation.

A cause and effect diagram is shown below for the sources of variation which might affect drive time to work. NOTE: Some of these causes are real and affect drive time to work, but some are just measurement error, but still does affect the resulting numbers.

Variation Effecting Drive Time to Work

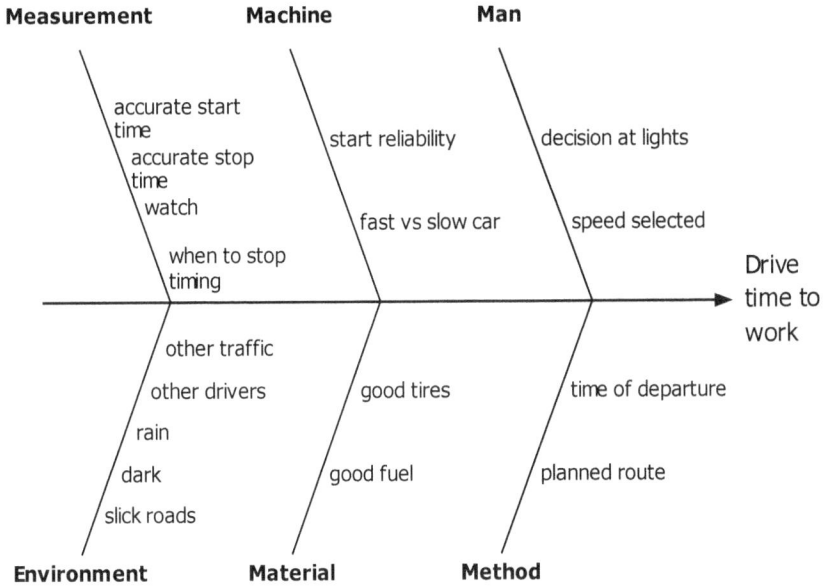

These diagrams are also known as fishbone diagrams or Ishikawa[1] diagrams.

Frequently these cause and effect diagrams are created in a group setting. If there is difficulty getting people to participate ... then a common method to encourage partici- pation is to create a negative cause and effect diagram.

People frequently generate more excitement in focus on the negative; thus, in this ex- ample, the focus would be how to make the drive time longer. Responses would be similar to above, but focused on lengthening drive time ... it would then be a simple matter to convert responses to causes of drive time variability.

---

[1] Dr. Kaoru Ishikawa, University of Tokyo, 1943.

Resolution ... Affects Measurement

A good common day comparison to understand resolution is thinking of a light switch at home versus a light with a dimmer dial or slide switch. The simple on-off switch has no resolution or "in-between adjustment" .... the light is either on or off. The dimmer switch may have some amount of increased resolution or finer increments or levels of "on". Now, think about a dimmer switch controlled by a dial that only turns ¼ turn versus a dimmer switch dial that can turn ¾ turn. The ¾ turn switch has more increments of control or better resolution.

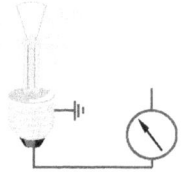

Many years ago there was a measurement system that used stones to weigh things. The stone method had very poor resolution and resulted in many things being called the same (no variation) when in fact there was considerable variation present.

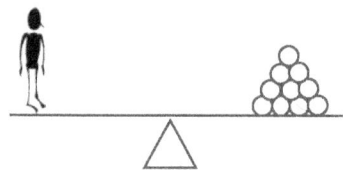

O =14 lbs ... Does this man weigh 140 lbs?

In modern days we have different units of measure and more accurate scales which can weigh in pounds, ounces, grams, etc.

If an old scale measures in pounds to the nearest ½ pound (0.5), then we get a scale that measures to the nearest 1/10 pound (0.1) ...we have better resolution, and can better identify variation. If we then get a scale that measures in grams to the nearest 1/1000 of a gram (0.001), we have even better resolution.

NOTE: There are other factors affecting the data accuracy such as:
1. measurement technique
2. calibration of the gage
3. accuracy of the gage (improved by calibration, but affected by quality of gage including resolution).
4. type of gage: automatic or manual
5. repeatability
6. reproducibility
7. other factors

Later we will discuss GR&R studies (repeatability and reproducibility) since it helps quantify how much variation is measurement error; thus, affecting overall variation. These GR&R studies are also known as MSE (measurement system evaluation).

Resolution in Graphs

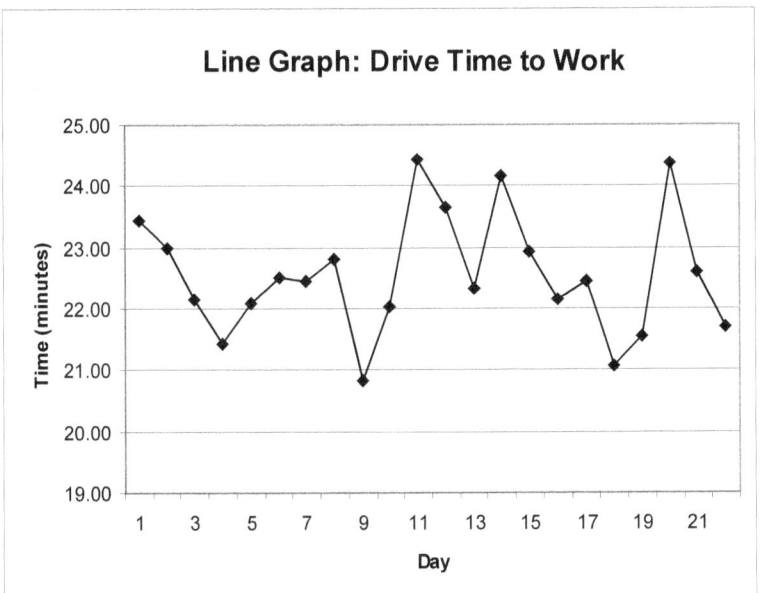

These two graphs depict the same data ... the difference is the y-axis scale. The top graph y-axis has a range of six minutes; whereas, the bottom graph y-axis has a range of 50 minutes.

The top graph has better resolution, and a better ability to display differences .... better discrimination. The bottom graph has a greater range, but does a poor job of displaying variation.

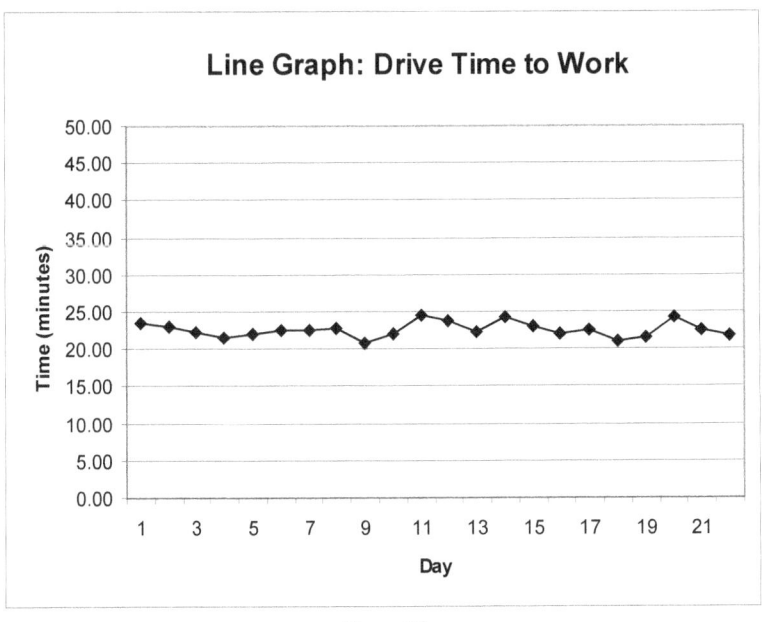

Resolution Also Affects Molding Process
(applied to molding machine selection)

During injection molding, the machine controls the linear movement of the screw at a set velocity ... if we inject the same volume of plastic over a longer length, we have better resolution of control (e.g. 7 inches vs an oversized press whereby stroke might only be 1 inch ... the ability to transfer using Decoupled Molding[SM] is better with longer stroke).

The bigger the barrel ID results in more volume (cm^3) of plastic being injected for each 0.001 inch of linear movement of the screw. The repeatability of limit switch sensitivity and response time of valves is only so good .... with that being constant, you will have better injection control with more stroke.

Of course, you still must engineer the system to achieve required pressure, screw recovery, fill rate, etc; thus, you must look at whole system.

When you have a 4 ounce barrel and use only 0.018 ounces ... you are using less than 10% of barrel capacity which results in this poor resolution of control as well as possible degradation from excessive residence time.

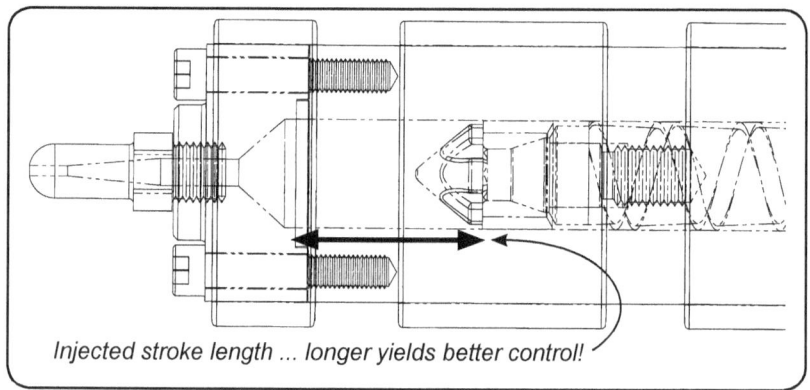

Injected stroke length ... longer yields better control!

DECOUPLED[SM] MOLDING ... OVERSIZED PRESSES

consider 4 parts weighing 0.514 grams in a 4 oz (30 mm screw) press:

$$\frac{0.514 \text{ gr}}{4 \text{ parts}} \times \frac{1 \text{ cc}}{1.06 \text{ gr}} \times \frac{1 \text{ mm}}{0.7068 \text{ cc}} \times \frac{0.03937 \text{ in}}{1 \text{ mm}} = \frac{0.0067 \text{ in}}{\text{part}}$$

30 mm screw diam = 15 mm radius = 1.5 cm radius

1mm STROKE = 1.5 cm² x 3.14 x 0.1 cm = 0.7068 cc

For each 0.0067 inches of screw ram travel, we would get 1 part ...

This is NOT good resolution of control to prevent overpacking

---

(Service Mark SM of RJG Associates, Inc)

Histogram

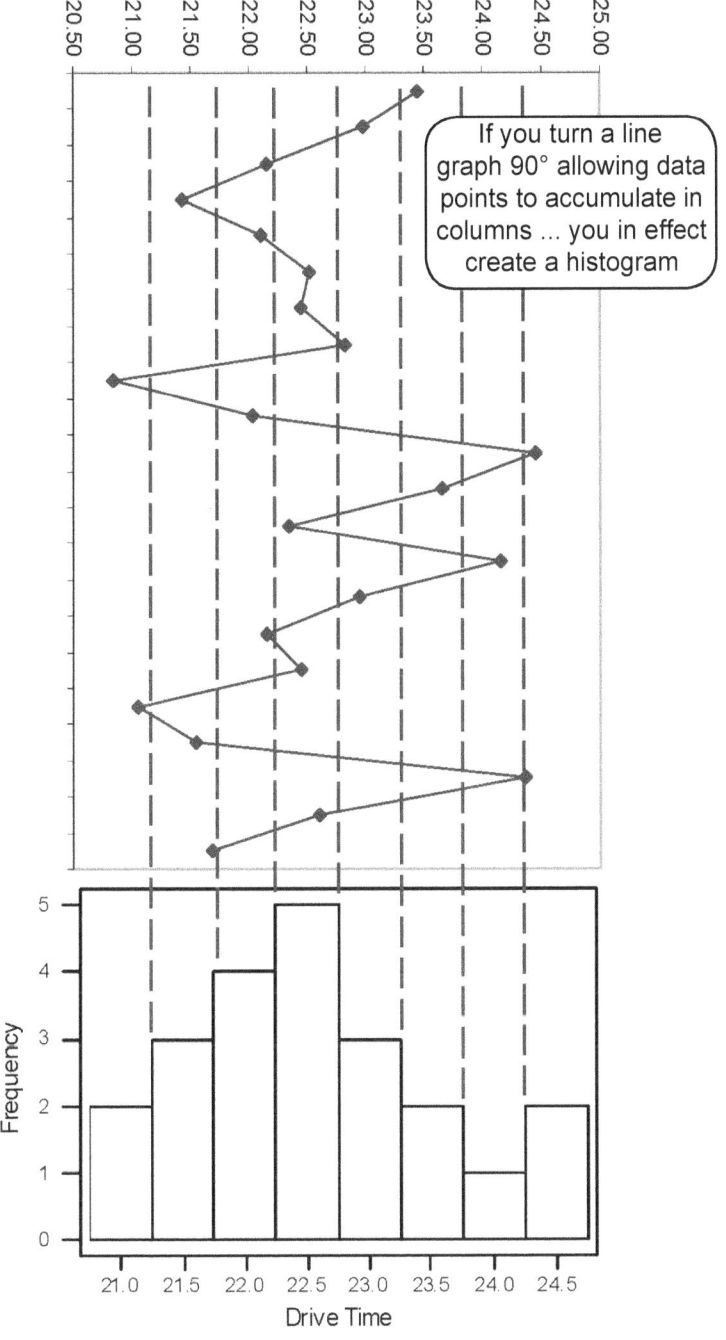

If you turn a line graph 90° allowing data points to accumulate in columns ... you in effect create a histogram

Histogram with Bell Curve

A histogram is a bar graph showing frequency of data points occurring in various intervals or groupings.

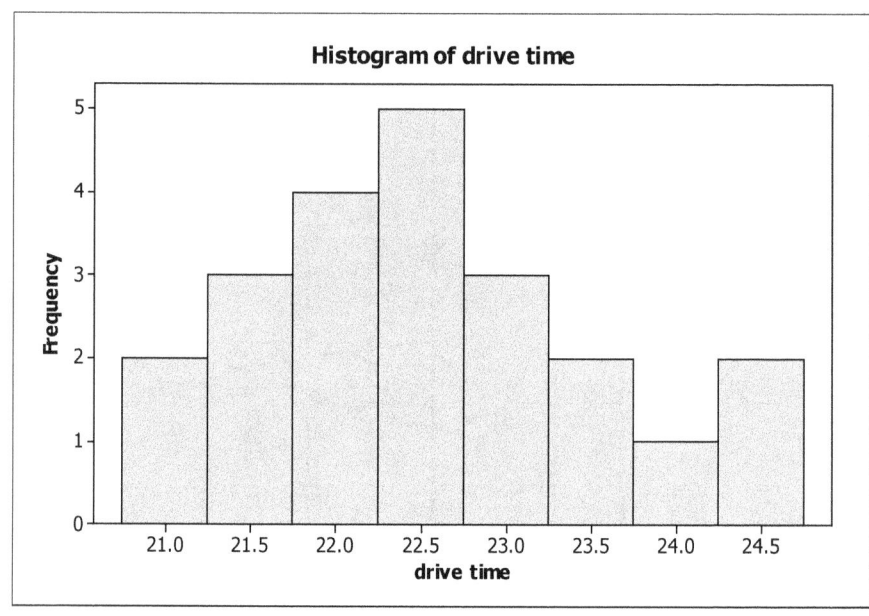

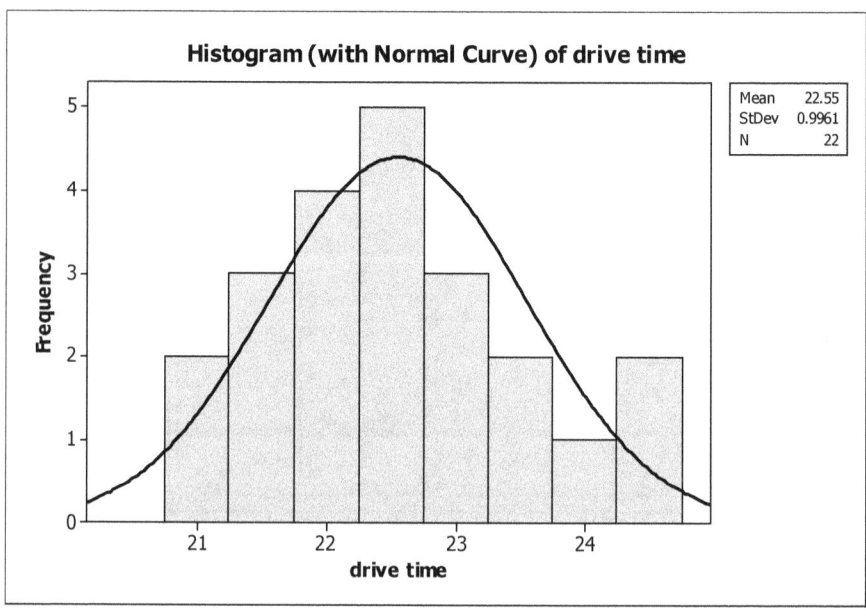

Line Graphs

On earlier pages we saw this line graph. Line graphs can be used to show variation and trends.

The x-axis on graph below is in chronological order; thus, we can see if drive time gets longer, shorter or just varies. In this example it just looks like normal variation.

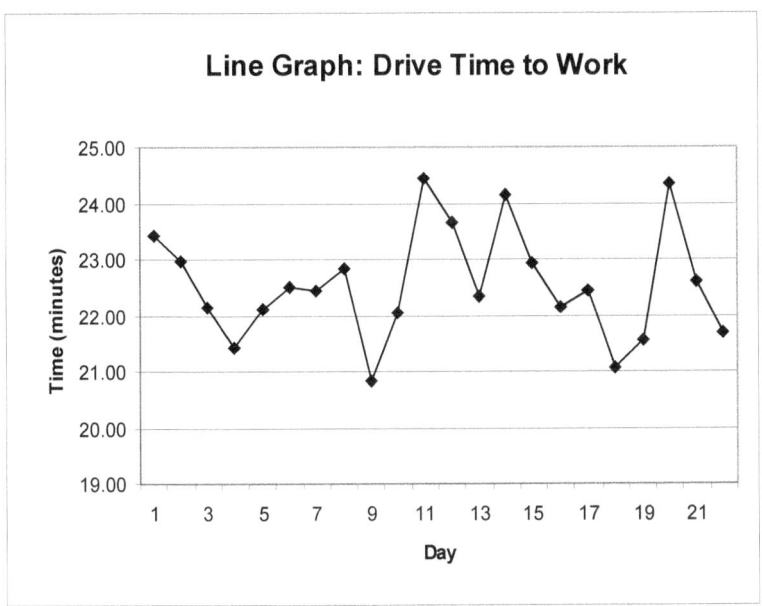

How can we conclude if it is just normal variation or excessive variation?

There are statistical techniques to analyze this data and then draw such a conclusion.

On next pages we will learn about standard deviation as a measure of variation.

We will also learn how to place an average and limits on the graph to see if the variation exceeds normal expected limits.

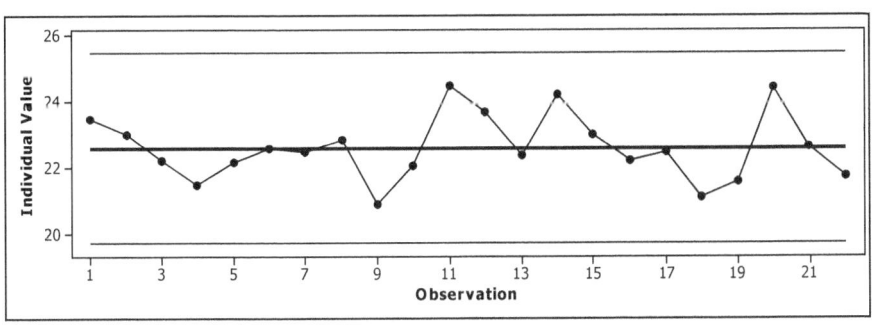

Mean, Max, Min, Range & Standard Deviation

In the drive time data set from the previous pages, we can analyze the numbers in various ways to draw conclusions about the data.

The average is simply the sum of all data divided by the count or n value - 22 entries.

$$\text{average} = \text{mean} = \frac{\sum\limits_{i=1}^{n} X_i}{n}$$

max = highest observed value

min = lowest observed value

range = max minus min

count = n = number of values

The standard deviation is a number which is derived from some slightly complicated math. The standard deviation value describes and actually quantifies the amount of variation; this important term is further described on next page.

standard deviation = s = sigma[1]

$$\sigma_{n-1} = \sqrt{\frac{\sum\limits_{i=1}^{n}(X_i - \overline{X})^2}{n-1}}$$

	Drive Time (min)
Day 1	23.44
Day 2	22.98
Day 3	22.16
Day 4	21.43
Day 5	22.10
Day 6	22.52
Day 7	22.45
Day 8	22.82
Day 9	**20.84**
Day 10	22.04
Day 11	**24.44**
Day 12	23.66
Day 13	22.34
Day 14	24.16
Day 15	22.94
Day 16	22.15
Day 17	22.44
Day 18	21.06
Day 19	21.56
Day 20	24.36
Day 21	22.61
Day 22	21.69
avg	22.55
max	24.44
min	20.84
range	3.60
count	22
std dev	0.9961

Since this formula looks overwhelming, please note the following: not as bad as it looks if we break up into pieces!

The standard deviation is the best value to quantify variation of most data sets ... sometimes range or min & max might be used.

---

[1] sigma can be a population sigma or a sample sigma; sample uses the n-1 divisor; whereas, population uses n as the divisor ... there is little difference between the two. In the manufacturing world, we seldom measure all parts in the population; thus, we will normally use sample sigma as shown in formula above.

Standard Deviation Calculated

In the table below, we will calculate standard deviation the hard way (manually).

	A	B	C	D	E	F	G	H	I	J
1		Drive Time		$(X-Xbar)^2$						
2		(min)								
3	Day 1	23.44		$(23.44-22.55)^2$	>>	0.89	X	0.89	=	0.7848
4	Day 2	22.98		$(22.98-22.55)^2$	>>	0.43	X	0.43	=	0.1814
5	Day 3	22.16		$(22.16-22.55)^2$	>>	-0.39	X	-0.39	=	0.1553
6	Day 4	21.43		$(21.43-22.55)^2$	>>	-1.12	X	-1.12	=	1.2636
7	Day 5	22.10		$(22.10-22.55)^2$	>>	-0.45	X	-0.45	=	0.2062
8	Day 6	22.52		$(22.52-22.55)^2$	>>	-0.03	X	-0.03	=	0.0012
9	Day 7	22.45		$(22.45-22.55)^2$	>>	-0.10	X	-0.10	=	0.0108
10	Day 8	22.82		$(22.82-22.55)^2$	>>	0.27	X	0.27	=	0.0707
11	Day 9	20.84		$(20.84-22.55)^2$	>>	-1.71	X	-1.71	=	2.9381
12	Day 10	22.04		$(22.04-22.55)^2$	>>	-0.51	X	-0.51	=	0.2643
13	Day 11	24.44		$(24.44-22.55)^2$	>>	1.89	X	1.89	=	3.5567
14	Day 12	23.66		$(23.66-22.55)^2$	>>	1.11	X	1.11	=	1.2230
15	Day 13	22.34		$(22.34-22.55)^2$	>>	-0.21	X	-0.21	=	0.0458
16	Day 14	24.16		$(24.16-22.55)^2$	>>	1.61	X	1.61	=	2.5789
17	Day 15	22.94		$(22.94-22.55)^2$	>>	0.39	X	0.39	=	0.1489
18	Day 16	22.15		$(22.15-22.55)^2$	>>	-0.40	X	-0.40	=	0.1633
19	Day 17	22.44		$(22.44-22.55)^2$	>>	-0.11	X	-0.11	=	0.0130
20	Day 18	21.06		$(21.06-22.55)^2$	>>	-1.49	X	-1.49	=	2.2323
21	Day 19	21.56		$(21.56-22.55)^2$	>>	-0.99	X	-0.99	=	0.9882
22	Day 20	24.36		$(24.3-22.55)^2$	>>	1.81	X	1.81	=	3.2613
23	Day 21	22.61		$(22.61-22.55)^2$	>>	0.06	X	0.06	=	0.0031
24	Day 22	21.69		$(21.69-22.55)^2$	>>	-0.86	X	-0.86	=	0.7467
25	MEAN	22.55						sum	=	20.8377
26	n	22						n-1	=	21
27							sum : (n-1)		=	0.992273
28							sqrt (sum ÷ (n-1))		=	0.9961

$$\sigma_{n-1} = \sqrt{\frac{\sum_{i=1}^{n}(X_i - \overline{X})^2}{n-1}}$$

In the calculation above, we used the formula for standard deviation and calculated 0.9961 as the std dev for the drive time data set. It took too much time and effort to do this. Fortunately, we will never calculate standard deviation in this manner again .... there are various computer programs or calculators or spreadsheet functions which can do this for us! We could have easily typed in the spreadsheet function: =STDEV(B3:B24) and gotten the answer of 0.9961.

Central Limit Theorem

Variation is inevitable and will manifest itself in one of two forms:

1. **Common cause** variation which will exist even if the process is in statistical control because it is due to random chance ... this type variation is predictable using the power of basic statistical math.

2. **Special cause** variation is not predictable; it is typically caused by some root cause which can be identified given sufficient investigation ... this type variation will violate the statistical control limits which allow only for common cause variation.

A process that is in control exhibiting only common cause variation will have data distributed in what is called a normal distribution curve whereby most data is clustered around a central mean. As data gets farther from the mean, it's frequency or number of occurrences will be less. The mean and standard deviation calculations will identify the shape of this bell shaped, normal distribution curve. The mean is the average of all data in the sample – center of the curve, and standard deviation will describe the amount of variation present in the sample – width of curve (± 3 to 6 std devs).

A cause and effect diagram can be made to better understand the many sources of variation. Some special cause variation may occur so frequently and be determined as uncontrollable. In this case, this special cause variation may be reclassified as common cause variation. This concept must be considered when planning a sub-grouping strategy for control charts (discussed more later).

The area under various portions of the curve are as listed in table below. In the ± three sigma chart shown: 99.73% of the parts would conform to this normal variation or common caused variation. Only three (2.7) times out of a thousand would variation fall beyond these three sigma expectations. Since 0.27% (0.0027) can still be an excessive number to some, many manufacturers today plan their capabilities around ± 4 sigma or even ± 6 sigma to yield less defects.

The next page will discuss capability in further detail. Later pages will compare Cpk vs Z score as a method to describe capability. The real purpose of a capability index or Z score is to quantify the risk of discrepant product being produced and sent to the customer. We will see later how to convert these values into ppm (parts per million) defective.

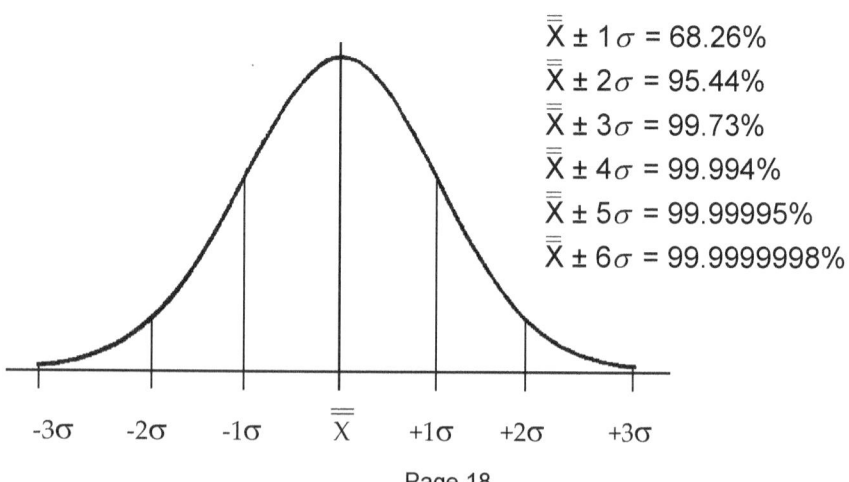

$$\bar{\bar{X}} \pm 1\sigma = 68.26\%$$
$$\bar{\bar{X}} \pm 2\sigma = 95.44\%$$
$$\bar{\bar{X}} \pm 3\sigma = 99.73\%$$
$$\bar{\bar{X}} \pm 4\sigma = 99.994\%$$
$$\bar{\bar{X}} \pm 5\sigma = 99.99995\%$$
$$\bar{\bar{X}} \pm 6\sigma = 99.9999998\%$$

-3σ -2σ -1σ $\bar{\bar{X}}$ $+1\sigma$ $+2\sigma$ $+3\sigma$

How Much Variation is Acceptable?

Variation is present in everything we do; including injection molding.

Variation is a "common" natural occurrence.

Variation exists when something is slightly different from another of the same type ... not the same.

- Natural variation – people who are different heights, weights, eye color, hair color, ambient temperatures, etc
- Logistical variation – shipping times, transit times, etc
- Dimensional variation – different lengths, widths, diameters, etc
- Defect type variation – flash, voids, sinks, FM, ECR, etc

There are techniques or methods which can be used to quantify variation known as SPC – statistical process control (looks at the voice of the process ... data from the output). SPC uses the power of statistics to define the acceptable limits of variability ... without this, we would not know when variation is normal versus excessive.

There are math formulas to determine how much variation should be considered normal or "common". When the variation exceeds this common threshold value, it is considered "special".

Benefits of SPC:

- In the manufacturing world: reduce risk of producing discrepant product.
- Identify special caused variation that may create the risk of producing scrap:
  - supplier issues such as resin
  - equipment issues
  - tooling issues
  - people, procedures[1] or system issues, etc
- Increase customer confidence in supplied products; respond to customer demands to show evidence of effective SPC

---

[1] Do not underestimate the importance of systems and procedures and their effect on SPC. Frequently as organizations implement SPC, they discover that increased attention to documentation and procedures is needed ... such as change control and notification. If the process is deliberately changed to solve a problem or as part of continuous improvement, clearly the control limits may be violated and require recalculation.

It is also beneficial to push SPC up to the supplier level. SPC will identify sources of variation at the manufacturing plant, but purchased materials may also introduce variation ... this variation should also be reduced to achieve maximum success.

Estimated Standard Deviation

We have been using line graphs and control charts long before there were calculators and computers to help us compute standard deviation. The calculation seen on page 17 would have been much too complicated to perform on a frequent basis to derive the measure of variation (standard deviation).

We will use this complex measure of variation (standard deviation) to derive our control limits ... or limits of acceptable variation (the lines indicating normal range of variation). See graph below (from Minitab®) ... there are now control limits: UCL & LCL.

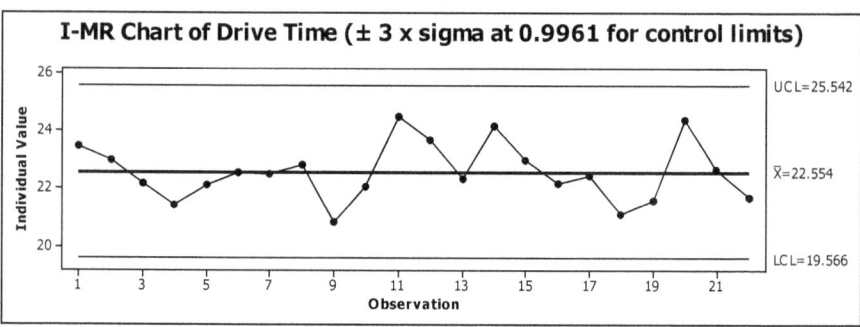

These control limits are the mean plus three standard deviations up and minus three standard deviations down (typ shown as ± 3 σ limits).

Since this standard deviation was so complicated to compute, a technique was developed a long time ago to estimate the standard deviation.

The range of data is also an indicator of variation. This range calculation is a simple subtraction (maximum minus minimum), but max what minus min what? What data do we look at to get this max and min?

In order to estimate the standard deviation, we typically create another chart below the main value chart. The main chart or top chart will list the plotted values, and is called a X-bar chart (in chart above it is called individual chart because we are plotting just one point (one drive time to work ea day), but if the one point is from an average of 2,3,4,5 points then it is called a X-bar chart .... same chart either way: average of one or average of 2,3,4,5 or more).

A second chart is used to estimate the standard deviation and is called a range chart or a moving range chart. The whole purpose of the range or moving range chart is to plot a series of range values to then get an average range called r-bar. If we are averaging 2,3,4,5+ as stated above, we can find that max and min and calculate a range. If we plot a single data point (like this example), then we have to get creative to calculate the range by subtracting previous single value from current single value to get a range; henceforth a moving range chart; we will use r-bar to estimate the standard deviation – the r-bar on next page is 1.080, but this is better derived on page 23. The purpose of the bottom chart (R or MR) is to derive the r-bar and then easily compute the estimated standard deviation .... aka sigma hat which is derived by dividing r bar by d2 (d2 also written as d_2 is a statistical unbiasing constant which adjusts the calculated value in attempt to best reflect the sample size ... i.e. n = 2, 3, 4, 5, etc).

Estimated Standard Deviation (continued)

The graph on previous page used ± 3 sigma limits for the drive time data set ... same graph also relisted below (± 3 true sigma, NOT estimated sigma).

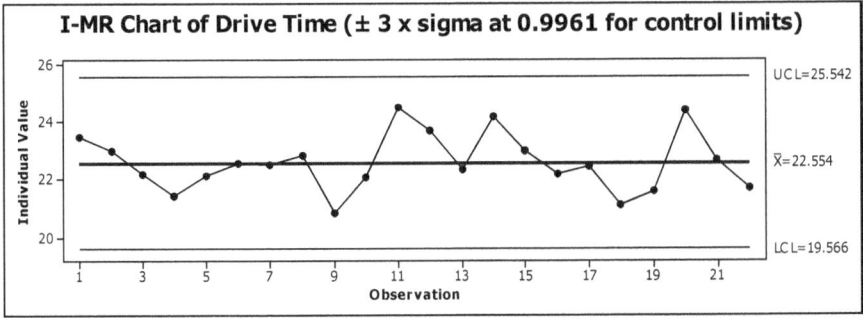

The same data is listed below, but now with slightly different control limits based on ± 3 estimated sigma limits ... note how similar the control limits are in proximity to the data which shows how effective estimated sigma can be. The standard deviation (sigma) was calculated earlier as 0.9961 ... the estimated sigma is MR ÷ d2 or 1.080 divided by 1.128 which equals 0.9579.

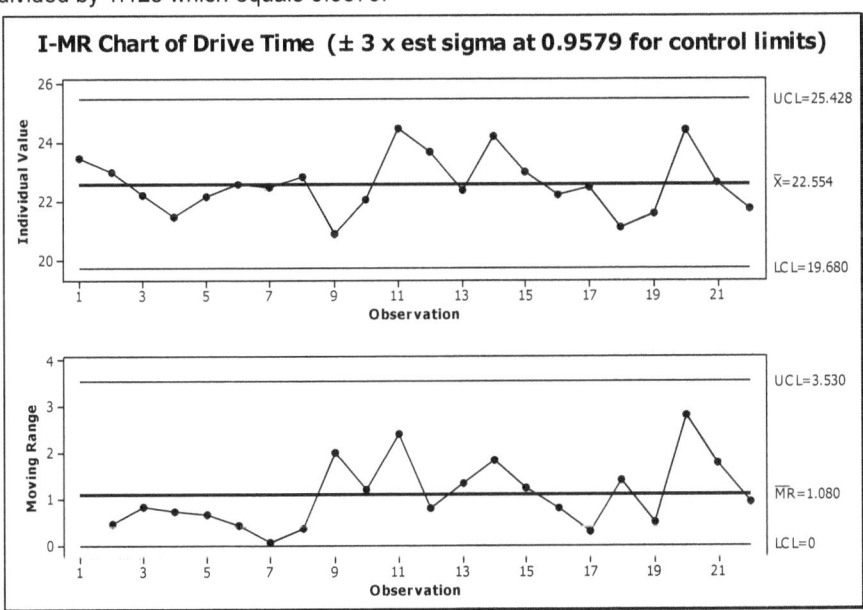

While the estimated sigma can be very accurate or accurate enough, it can also be very inaccurate; thus, take care when setting up control charts and selecting a sub-grouping strategy and determining the standard deviation or estimate thereof.

In future pages, it will be shown whereby data with large shifts and drifts which happen in small increments can result in a largely incorrect estimated sigma – the estimated sigma based on short term ranges can grossly under estimate the overall variation.

Graph of Drive Time with MR chart above to yield r-bar value. Graphs this page made using MINITAB® v. 16 from Minitab Inc.

Estimated Standard Deviation (continued)

We will designate an estimated standard deviation as sigma-hat. Sigma-hat is derived by taking the R-bar from range chart (the lower chart from an X-bar & R chart or X-bar MR chart), then divide by d2 .... d2 is a constant[1] read from a table of constants (table below). The d2 constant is based on sample size. In this case, not the 22 different drive times, but rather n=1 (becomes n=2) for drive time example. In an X-bar & MR chart (or data set) like previous and next page, the range comes from the difference between successive days (i.e. day 2 minus day 1; day 3 minus day 2, day 4 minus day 3, etc); thus, n=1 actually becomes n=2). The term "n" is typically used to describe the subgroup sample size, not the number of observations, but n is often used to describe both – can be confusing.

A X-bar & R chart will also have an R-bar regardless of n value .... n can be 2,3,4,5 or more (typical is 3,4 or 5).

In a X-bar & MR chart: the MR-bar is the average of all the Rs or ranges (see example calculation on next page).

In a X-bar & R chart: the R-bar is the average of all the Rs or ranges.

The reason for this estimated standard deviation comes from the difficulty in calculating the actual std dev (seen on earlier pages) in a time before powerful calculators and computers ... it does beg the question, why ever use estimated std dev today?

Xbar & R Charts ... Xbar/Moving Range Charts				
Sample Size				
n	A_2	D_3	D_4	d_2
2	1.880	---	3.267	1.128
3	1.023	---	2.574	1.693
4	0.729	---	2.282	2.059
5	0.577	---	2.114	2.326
6	0.483	---	2.004	2.534
7	0.419	0.076	1.924	2.704
8	0.373	0.136	1.864	2.847
9	0.337	0.184	1.816	2.970
10	0.308	0.223	1.777	3.078

---

[1] These constants are called unbiasing constants ... the differences attempt to unbias any results caused by larger or smaller sample sizes. A larger subgroup of n would likely contain more variation; thus, different constants.

See also page 33 for further discussion of basis for A_2 and D_2.

Est. Standard Deviation – Drive Time Example

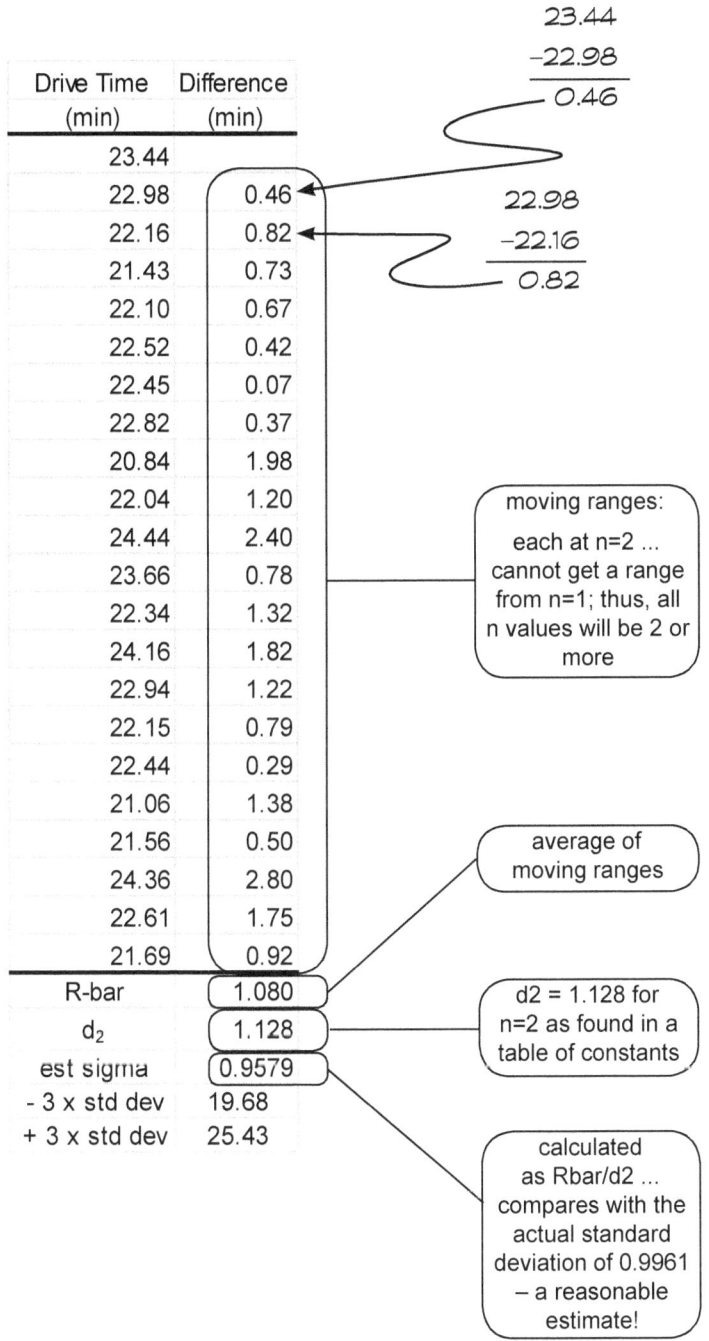

$$\begin{array}{r} 23.44 \\ -22.98 \\ \hline 0.46 \end{array}$$

Drive Time (min)	Difference (min)
23.44	
22.98	0.46
22.16	0.82
21.43	0.73
22.10	0.67
22.52	0.42
22.45	0.07
22.82	0.37
20.84	1.98
22.04	1.20
24.44	2.40
23.66	0.78
22.34	1.32
24.16	1.82
22.94	1.22
22.15	0.79
22.44	0.29
21.06	1.38
21.56	0.50
24.36	2.80
22.61	1.75
21.69	0.92
R-bar	1.080
d_2	1.128
est sigma	0.9579
- 3 x std dev	19.68
+ 3 x std dev	25.43

$$\begin{array}{r} 22.98 \\ -22.16 \\ \hline 0.82 \end{array}$$

moving ranges: each at n=2 ... cannot get a range from n=1; thus, all n values will be 2 or more

average of moving ranges

d_2 = 1.128 for n=2 as found in a table of constants

calculated as Rbar/d2 ... compares with the actual standard deviation of 0.9961 – a reasonable estimate!

Standard Deviation Summary for Drive Time

A process that is in control exhibiting only common cause variation will have data distributed in what is called a normal distribution curve whereby most data is clustered around a central mean. As data gets farther from the mean, it's frequency or number of occurrences will be less. The mean and standard deviation calculations will identify the shape of this bell shaped, normal distribution curve.

The relationship between the mean or average and the standard deviation is listed in the graphic below. By definition, 68.26% of the data will fall within ± 1 sigma from the mean; 95% will fall between ± 2 sigma and 99.73% will fall between ± 3 sigma.

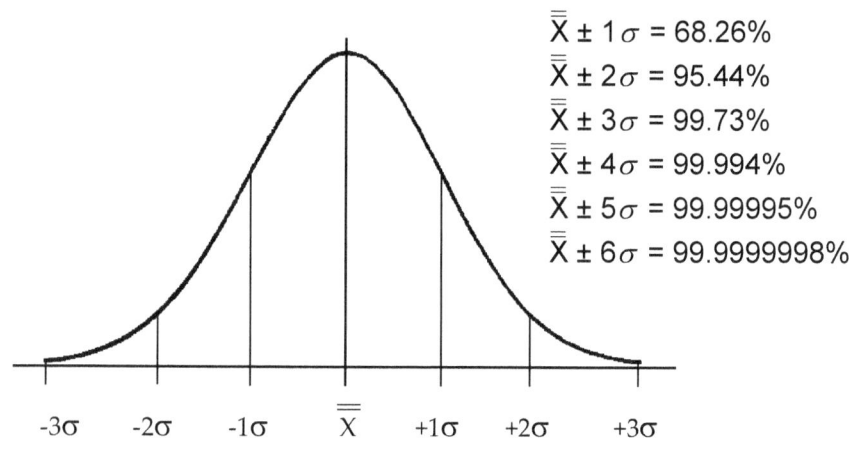

$$\bar{\bar{X}} \pm 1\sigma = 68.26\%$$
$$\bar{\bar{X}} \pm 2\sigma = 95.44\%$$
$$\bar{\bar{X}} \pm 3\sigma = 99.73\%$$
$$\bar{\bar{X}} \pm 4\sigma = 99.994\%$$
$$\bar{\bar{X}} \pm 5\sigma = 99.99995\%$$
$$\bar{\bar{X}} \pm 6\sigma = 99.9999998\%$$

-3σ   -2σ   -1σ $\bar{\bar{X}}$ +1σ   +2σ   +3σ

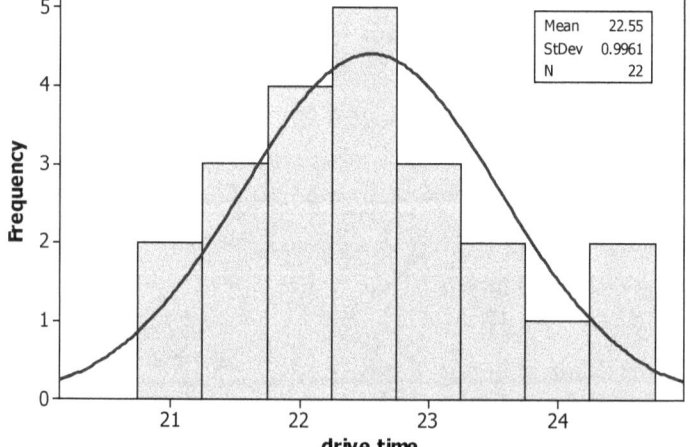

Histogram of Drive Time
Normal

Mean	22.55
StDev	0.9961
N	22

Range of what should be considered normal drive time (we should always leave at least 25.54 minutes before work start time …more to allow for special circumstances).

19.57   mean - 3 x std dev

25.54   mean + 3 x std dev

Common vs. Special Cause Variation

Common variation is natural and mostly unavoidable and will be discussed in more detail on later pages. When the variation goes beyond the levels established for "common" variation, then it is "special".

In our drive time example, we can list some examples of common cause variation vs special cause variation.

Variation	
Common	Special
hitting more or less red lights	have flat tire
driving slightly faster/slower	stop for gas
moderate changes in traffic	accident causing heavy traffic
parking location in lot	detours

In our data collection period of 22 days, if there was only common cause variation (no flat tires, etc), then we can predict the future drive times as typically taking between 19.57 minutes and 25.54 minutes. This is computed by taking the mean and adding 3 sigma to get upper end of expected drive time .... and also subtracting 3 sigma from the drive time to get the fastest drive time to be expected.

$$22.55 + (3 \times 0.9961) = 25.54 \text{ minutes}$$

$$22.55 - (3 \times 0.9961) = 19.57 \text{ minutes}$$

If we decide that we want to sleep longer, and cannot tolerate the drive times of 19.57 to 25.54 minutes ... can we change these numbers?

Yes – if we take some corrective actions to reduce the drive time such as:
1. move closer to work
2. find a better route
3. drive during periods of less traffic; leave earlier
4. etc

There are frequently corrective actions which can either shift the mean (average drive time) or reduce the variation ... these may be easy or difficult; analysis may be required.

Test Data for Normal Distribution

I-MR Chart of drive time ... flat tire on fifth day

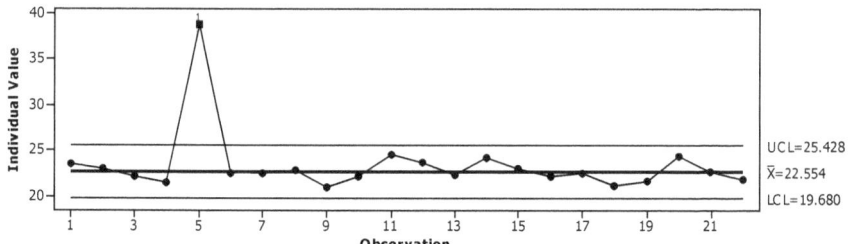

When we establish the control limits, our data should be normally distributed, and without special cause variation as seen on data set above where a flat tire occurred.

In Minitab, we can test for normal distribution whereby we want a p-value from such a test to be greater than 0.05. The flat tire month would not be a good month to set control limits because we have a major outlier – such data is NOT normally distributed.

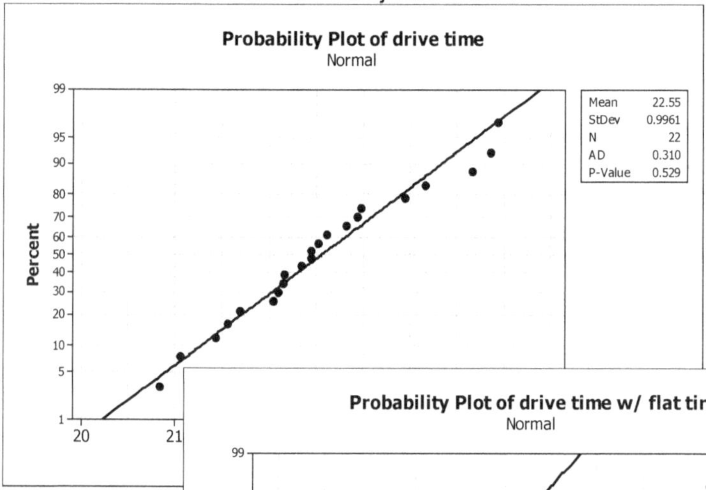

Probability Plot of drive time
Normal

Mean	22.55
StDev	0.9961
N	22
AD	0.310
P-Value	0.529

Always test data to verify it is normally distributed before setting control limits!

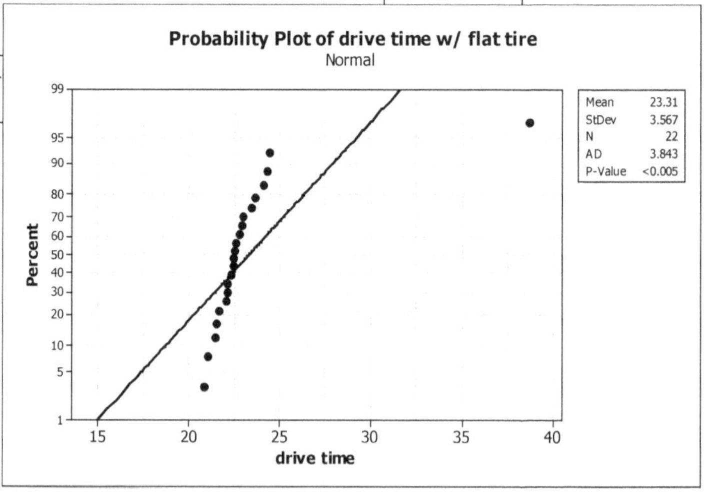

Probability Plot of drive time w/ flat tire
Normal

Mean	23.31
StDev	3.567
N	22
AD	3.843
P-Value	<0.005

Use Anderson-Darling test for normality: ignore the AD number; look for p value > 0.05 to indicate a normal ditstribution. In Minitab, Select: Stat, Basic Statistics, Normality Test, select the Variable to test ... select: Anderson-Darling, click OK.

Graphs made using MINITAB® from Minitab Inc.

Rules for Control Charts

1. Select a measurement that is the voice of the process (or indicative of overall product quality).
2. Investigate the measurement system to make sure it is capable: GR&R <10%, if possible, to insure the variation is from the manufacturing process and not from the measurement process. Typical MSE numbers may be higher than these guidelines ... see later pages on this subject.
3. While collecting data to establish limits, the data should only have random variation – no special caused variation caused by assignable causes ... we would not use the data having a flat tire to calculate control limits. As stated on previous page, you should perform a test for normality to verify that the data distribution is normal.
4. The data should be time ordered – in chronological order.
5. When sub-grouping, the subgroups should be rational ... meaning the parts and resulting measurements should be independent[1] from each other. NOTE: This can be a problem in injection molding because a primary source of variation is sometimes the resin and there are typically 3-5 shots in the barrel becoming mixed (thus, the sub-groups are not independent since the resin variation is being mixed for 3-5 consecutive shots) ... this results in little variation between successive shots. For this reason, it is often preferred to use n=1 (math from 2 values) and a Xbar-MR chart for injection molding. If you use a X-bar & R chart with n=3,4,5; then be sure not to collect successive shots.

   NOTE: This important point will be listed multiple times in this book!

6. The control limits should be <u>statistically derived</u>: do not use specification limits as control limits.
7. Perform root cause analysis for points out of control. With ± 3σ limits, you will get false alarms approximately 0.27% of the time ... one cannot guess when these false alarms happen; thus, you need to investigate OOC conditions. Since item 7 is very important, you need to pay close attention to item 6 and how you calculate control limits. You do not want a situation whereby OOC is normal. Research also response plans for alarms and/or control limit violations.

---

[1] Autocorrelation can exist when the subgroups are not independent. Autocorrelation conditions frequently result in control limits that are much too tight to be meaningful. Autocorrelation exists when each observation is predictable by the observation immediately before it. This can occur when successive molded shots are taken because the molding machine's heated barrel, that plasticizes the resin, has pre-mixed the resin. This results in each shot having essentially the same blend ... resin is a significant source of variation in injection molding resulting from the resin suppliers process to make.

Problems with Est Std Dev (estimated sigma)

Note: The data below exhibits drifting data ... this data (not real data) is greatly exaggerated to make the point! ... but shifts and drifts do occur.

The data is exaggerated to demonstrate the potential difference between actual standard deviation and an estimated standard deviation. The n value is 2.

The control limits for a X-bar & R chart are typically $\pm A_2$ x R-bar (A_2 is another constant based on n value).

Control limits = $\pm A_2$ x Rbar

X-bar UCL = 10.4438

X-bar LCL = 10.4367

UCL - LCL = 0.0071

A_2 = 1.88

Rbar = 0.001881

Equivalent estimated sigma = (1.88 x 0.001881) ÷ 3

Equivalent estimated sigma = 0.00118 ... derived from A_2 x Rbar ÷ 3

Estimated sigma = (R-bar ÷ d2) = 0.00167

Observations = 270

Actual sigma = 0.02577 ... 15 times higher than the estimated sigma; not close at all!

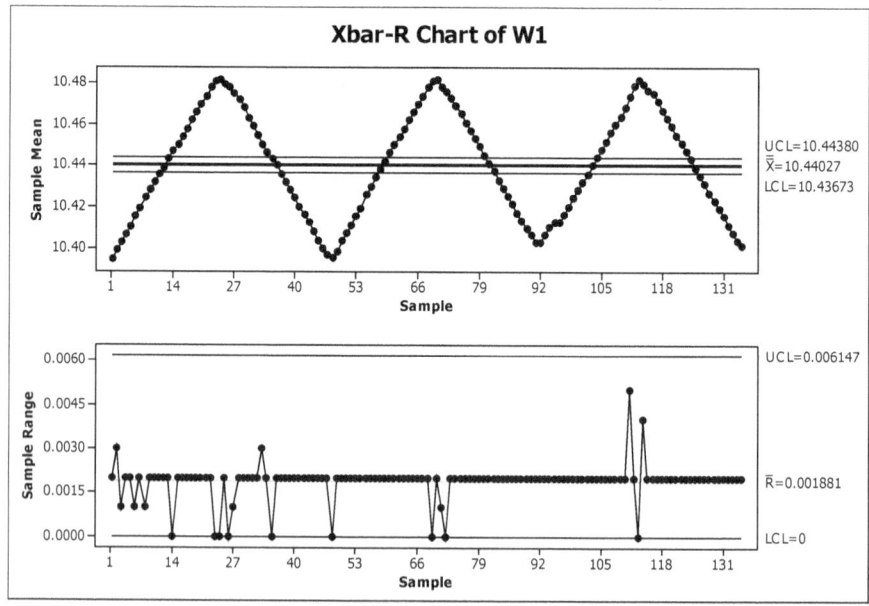

If the data drifts, AND drifts in small increments as shown, the estimated sigma (standard deviation) will be falsely low resulting in very tight control limits. An estimated standard deviation only looks at 2, 3, 4 or 5 parts typically (especially for an X-bar & R chart) ... and frequently these are consecutive parts.

Graphs made using MINITAB® from Minitab Inc.

Injection Molding and Rational Subgroups

- Good SPC control charting requires "rational" subgrouping strategies …. meaning individual part measurements are independent from each other and only effected by random variation.
- In injection molding processes with n=2,3,4,5 as a subgrouping strategy … there will be little variation between successive molded parts because there are typically 2-4 or more shots in the barrel which has become mixed whereby resin and melt temperature changes are no longer independent.
- The best way of achieving independent subgroups is with an I-MR chart … most injection molding variation is longer term in nature > 1-2 hrs …. up to multiple days or weeks. If a traditional Xbar & R chart is used: do not use consecutive shots that are less than a minute apart.
- A X-Bar & MR chart permits some longer term variation into the lower chart … remember that the R-bar is used to compute the X-bar chart control limits.
- Another way of achieving independent subgroups would be to collect the n values at different points of time (e.g. n=3 whereby 1 part from each hour). This will still not reflect the longer term resin variation or variation caused by mold cleanliness or venting characteristic changes.

Technically, a resin lot change that has different resin properties (molecular weight, weight distribution, additives, etc) is considered special caused variation, and might cause an out of control condition.

The dilemma for molders is: there will be many resin lot changes in the course of a year … do we want to recompute control limits at every lot change … even if it is special cause variation; it is a common occurrence in injection molding … even if we were willing to recompute limits that often, there is time required to establish a new basis (a new data set) for computing new limits.

This point regarding n=2 or n=3 is relisted because so many molding plants use n=2 or n=3 with poor control charting results. These results are particularly bad if the thermoplastic resin used includes a propensity for variation – common in many thermoplastics.

Such high variation thermoplastics include olefins like polyethylene or polypropylene. The nature of these thermoplastics include differences in molecular chain length (different molecular weight) and differences in additives that effect crystallization rates which greatly effect shrink rates and ultimate shrinkage. The molecular weight greatly effects viscosity and fill rates, packing rates, pressure drop, etc.

Some thermoplastics are very consistent and will not exhibit problems with traditional Xbar & R charts using n=3 … but be aware that these sub-groups still have the resin, additives at the press such as color or regrind, and melt temperature blended in the barrel making variation from these sources no longer distinguishable from each other.

Example: X Bar and R Control Charts

HOUR	1	2	3	4	5	6	7	8	9	10	11	12	13	14	15	16	17	18	19	20	21	22	23	24
	76	74	77	79	75	74	72	77	75	76	74	73	77	72	74	76	76	73	71	78	79	74	73	76
	74	72	73	73	74	74	73	77	76	74	72	74	75	76	71	71	72	77	74	75	77	73	71	69
	69	71	78	74	74	73	77	71	74	79	72	73	76	71	77	76	78	72	74	74	72	77	65	76
	72	69	71	72	74	74	75	74	76	77	73	73	71	73	72	73	74	75	75	76	77	73	76	74
	74	74	75	76	76	74	73	73	71	71	77	71	75	77	74	75	73	77	77	75	72	71	71	71
Xbar =	73	72	74.8	74.8	74.6	73.8	74	74.4	74.4	75.4	73.6	72.8	74.8	73.8	73.6	74.2	74.6	74.8	74.2	75.6	75.4	73.6	71.2	73.2
Range =	7	5	7	7	2	1	5	6	5	8	5	3	6	6	6	5	6	5	6	4	7	6	11	7

$\bar{\bar{X}}$ = avg of $\bar{X}$s = Grand Avg or Grand mean = 74.025 = avg of all 120 "X" values

$\bar{R}$ = avg of Rs = 5.666

look up values for A_2, D_4 & D_3 below in appropriate table of constants

X bar chart; upper control limit (UCL) = $\bar{\bar{X}} + A_2\bar{R}$ = 77.29 (A_2 = 0.577 for n=5)

X bar chart; lower control limit (LCL) = $\bar{\bar{X}} - A_2\bar{R}$ = 70.76 (A_2 = 0.577 for n=5)

Range chart ... upper control limit (UCL) = $D_4\bar{R}$ = 11.98 (D_4 = 2.114 for n=5)

Range chart ... lower control limit (LCL) = $D_3\bar{R}$ = 0 (D_3 = 0 for n=5)

- D_3, D_4 & A_2 are found from a table of constants for variables control charts using n = 5 (5 ea hr).
- After comparing the actual values to the control limits, it can be seen the process is in control.
- It is recommended that the above data be plotted graphically to better display the data trends, etc.

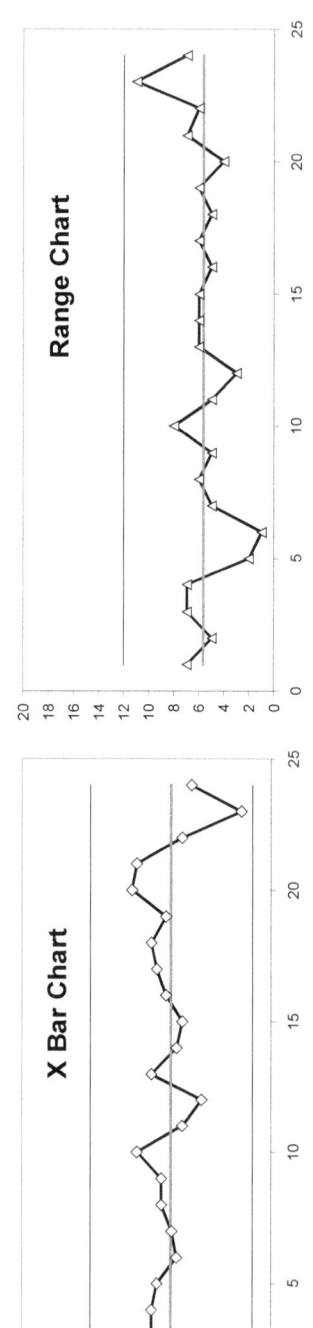

X Bar Chart

Range Chart

X-bar & R vs. X-Bar & MR Chart

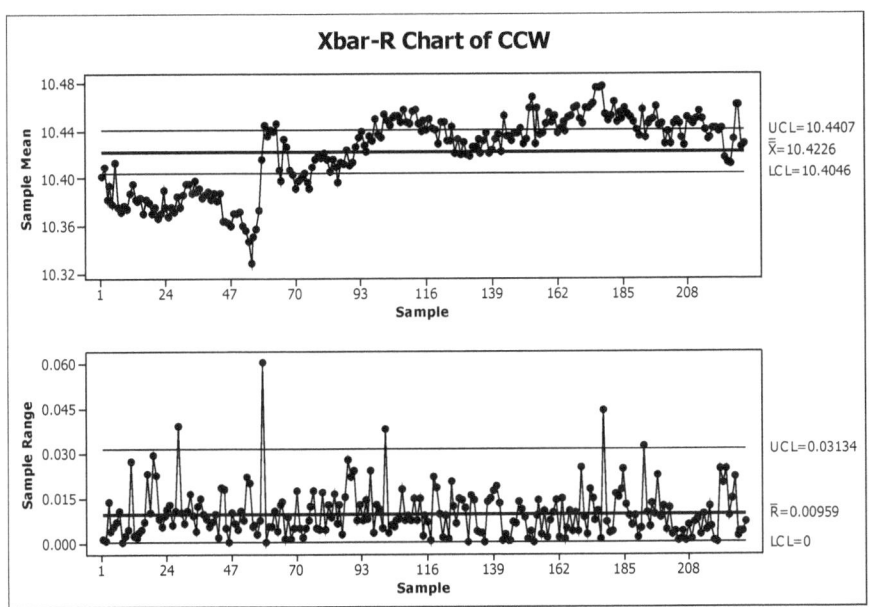

In the X-bar & R chart above, the subgrouping strategy was to take 2 samples (n=2), then plot the range of these 2 (could be 3,4,5, etc) in the range chart. The average of the 2 is plotted in the X-bar chart. Control limits are based on $\pm A_2$ x R-bar.

In the I-MR chart below .... all points are plotted with the limits used being ± 3 sigma computed from only the data after water problem fixed (approx point 110).

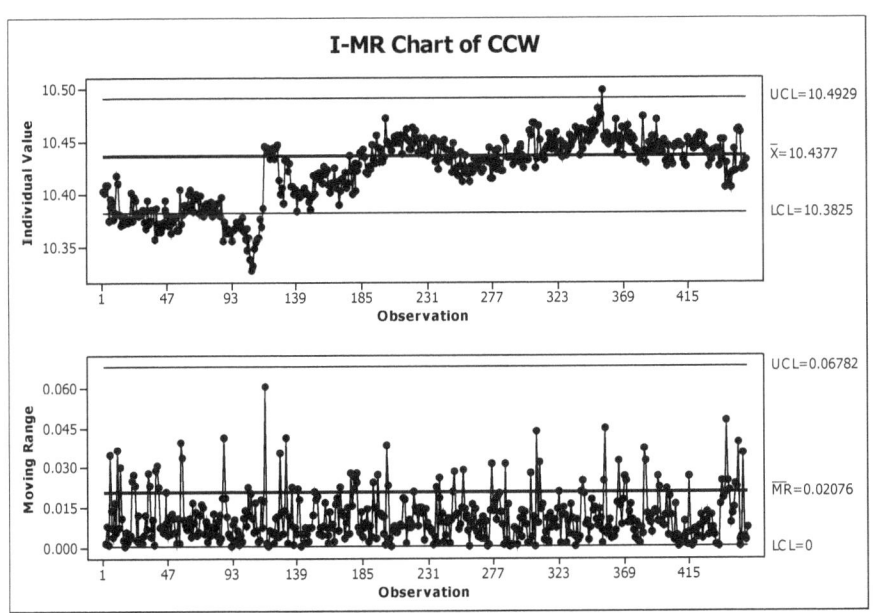

Alternate Methods for Calculating Control Limits

As mentioned on previous pages, <u>the range chart in normal control charting drives the control limits for the Xbar control charts.</u> The only problem with this approach for injection molding is that typically (not required), the sub grouping strategy uses consecutive shots which co-mingles the resin variation (and melt temperature variation) creating control limits that may be too tight to achieve compliance. The control chart is useless if out of control is the norm.

Alternative methods for computing control limits for Xbar chart include the following:
- ± 3 estimated sigma (aka 2.659 x Rbar) ... typically used in Xbar - MR chart
  Note: 2.659 comes from 3 ÷ 1.128 ... d_2 for n=2 is 1.128 ...
  3 x est sigma (Rbar ÷ d_2) = 2.659 x R-bar
- ± A_2 x Rbar ... A_2 based on the relationship between std dev for individuals vs std dev for averages ... see next page
- ± 3 calculated or true sigma
- ± 3 pooled sigma ... when all the data is in a single sample, pooled sigma and sample sigma are the same, but when precomputed sample sigmas (std dev) are calculated and merged AND these sigmas have different n values, then the following equation must be used to merge or pool the data and compute the composite standard deviation which gives proper weight to each contributing sigma

$$s_p = \sqrt{\frac{(n_1 - 1)s_1^2 + (n_2 - 1)s_2^2 + ...(n_k - 1)s_k^2}{n_1 + n_2 + ...n_k - k}}$$

The aforementioned is merely food for thought ... keeping in mind the main goal is to catch discrepant product and pass acceptable product. There is no single approach that is best for all scenarios ... it depends on how critical the product is with regards to it's performance requirements. Note the importance of deciding on a sub-grouping strategy ... such as 2 consecutive shots from an injection molding operation!

We are always trying to let the power of statistics, indicate the acceptable limits of variability ... when is variation too much or unacceptable for our manufacturing scenario and customer expectations?

IMPORTANT: We also want to be sure that the basis (data) used for computing control limits does NOT include special caused variation ... no data that includes assignable causes such as water problems, equipment problems, blend problems ... no data associated with any known problems!

Basis for A2 ... A2 Calculated ... D2 Discussion

Note: For each average: there may be 2 or more individuals located left and right (lower and higher) to make up the average ... this is important to understand since the averages (Rbar) may be used to estimate the standard deviation and resulting limits ... clearly the individuals will have a broader distribution.

The basis for A_2 in control limits is calculated below:

$$\sigma_x = \sigma_{\bar{x}} \sqrt{n}$$

3 sigma limits become:

$$3\sigma_x = 3\sigma_{\bar{x}} \sqrt{n}$$

substituting est sigma $\left(\bar{R}\big/ d_2 \right)$ for sigma (σ_x)

$$\frac{3 \left(\bar{R}\big/ d_2 \right)}{\sqrt{n}} = 3\sigma_{\bar{x}}$$

for n=2

$$\frac{3\big/ 1.128}{1.4142} \times \bar{R} = 1.881 \times \bar{R} = A_2 \times \bar{R} = \text{control limits for } 3\sigma_{\bar{x}}$$

The A2 value control limits in X-bar & R charts calculated below using Excel:

	A	B	C
1	n	d_2	A_2 (calc)
2	2	1.128	1.881
3	3	1.693	1.023
4	4	2.059	0.729
5	5	2.326	0.577
6	6	2.534	0.483
7	7	2.704	0.419
8	8	2.847	0.373
9	9	2.970	0.337
10	10	3.078	0.308

cell C2=(3/B2)/SQRT(A2)

Note: Beyond the scope of this booklet, but d2 can also be calculated or proven by creating 10,000 random numbers having a mean of zero and a standard deviation of one: use EXCEL function =NORMSINV(RAND()).

Then paste cell results as values in a separate column (paste special as values).

Then compute the moving range difference between each cell: use EXCEL function such as =MAX(F6:F7)-MIN(F6:F7) ... copy cells down

Then find Rbar of these 9999 moving ranges ... should get near 1.128

Control Charts and SPC

In many molding plants, the control chart is the basis for the plant's SPC effort at controlling the process and resulting quality. Often times, the control chart is used to track various dimensional requirements of the molded part. When we use control charts in this manner, we really are performing SQC (Statistical Quality Control) vs. SPC (Statistical Process Control). Both use the power of statistical math to define the acceptable limits of variability.

The most common technique is to use X bar and R charts which derives an estimated standard deviation by taking R bar and dividing by d_2 – a statistical constant which varies depending on the "n" value – number of samples in the subgroup. Long before the days of computers and better calculators there was a real need for an estimated standard deviation – that being simplified math. The X-bar charts for both SQC and SPC often use the classical statistical formula which takes the X double bar or grand average $\pm A_2$ times R bar to get upper and lower control limits.

$$\text{Control limits} = \overline{\overline{X}} \pm A_2 \overline{R}$$

SQC relies on part inspection which is typically a sampling plan. A sampling plan typically looks at a few samples (molded parts), and assumes that those parts represent the entire population for a given time period. The main disadvantage of the SQC approach relates to the time lag between sampling and measurement vs. when parts are produced. Some discrepant product may be saved which will not be discovered until the next sample is taken. This assumes the sampling is being done on a timely basis AND data reviewed to accomplish any control. This also assumes that the "control dimensions" selected are a valid "voice" for the process. The control dimensions are typically the few selected features measured when samples are taken. If the control dimensions are a valid "voice" for the process, then there would be a good correlation to other important dimensions – when the control dimension is good, so are other dimensions and when the control dimension is bad, so are other critical dimensions. Control charts will tell us when to leave process alone vs. when do we make adjustments – assuming an adjustment will correct the assignable cause. Sometimes we know the resin has shifted and this may be proven to be a statistically significant assignable cause, but if we know this happens and cannot control it we may decide to live with it in the short run. This decision assumes only control limits are violated and not engineering tolerances or specification limits. If it is only control limits, then we may choose to recalculate control limits at that time.

The specification limits are known as the voice of the customer.

The control limits are the voice of the process; ideally the control limits will fit inside the specification limits.

One significant challenge for effective control charting is a proper sub-grouping strategy. The sub-groups should only include normal or common causes of variation. Within sub-group variation will be plotted in the range chart. The range chart Rbar is used to calculate the Xbar chart control limits. Rbar $\div d_2$ is an estimate for standard deviation; thus, we will have statistically based control limits to signal when the process is out of control – special caused variation instead of common caused variation.

Control Charts (continued)

The operative word in control charts is control. Some shops gather such data to document what happened (a history chart).

A process that is in control permits the future to be predicted and uses the power of statistics to identify what is normal variation. The most classic error of control chart misuse is not calculating statistical control limits. Some companies compare the actual charted data to the specification limits. If data exceeds specification limits, then a process adjustment is made equal to distance from mean. The graph below shows the increased variation resulting from such a plan: the dark black line data is no adjustments vs data (gray solid line) with cumulative adjustments each time the spec limits (2.370-2.380) were exceeded ... the I-MR chart (next page spread) shows the original data to be in control; thus, not warranting an adjustment (the mean is near nominal). DO NOT MAKE <u>ROUTINE</u> ADJUSTMENTS TO A PROCESS IN CONTROL.

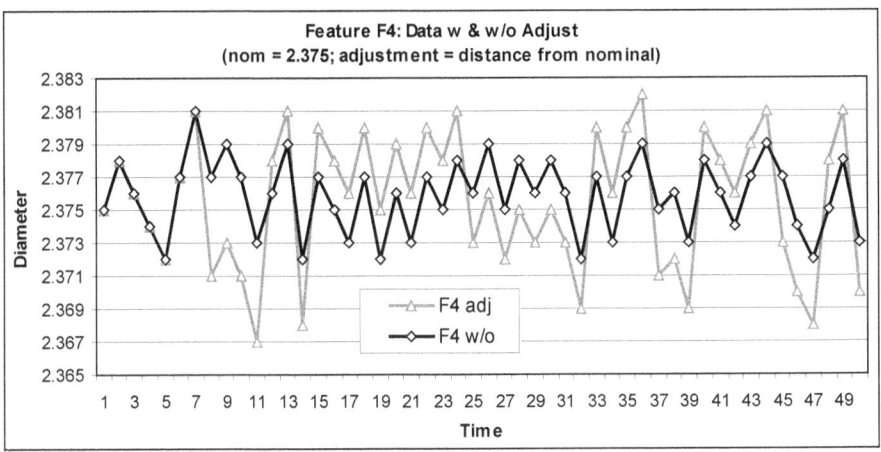

The line graph above is from Microsoft Excel®.

Control Charts Capability Analysis

Both charts have same spec limits; the top chart is from a process whereby each time spec limits were violated an adjustment was made equal to distance from nominal!

Process Capability Analysis for F4 adj

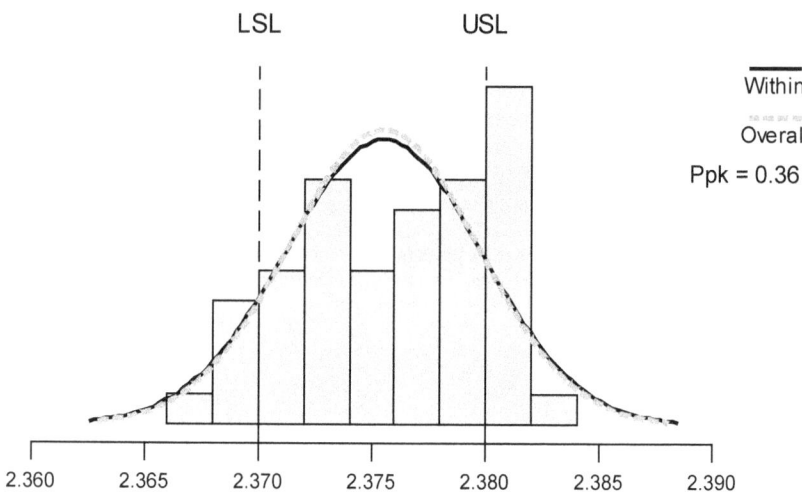

The graphical analysis below shows process capability with no adjustments (Ppk = 0.61); whereas, analysis above includes many adjustments (Ppk = 0.36) ... graphs from Minitab®. Note: Spec limits relative to data!

Too many adjustments can increase variation; do not over react to points OOC ... might be an outlier or measurement error, but do investigate and measure latest parts.

Process Capability Analysis for F4 w/o

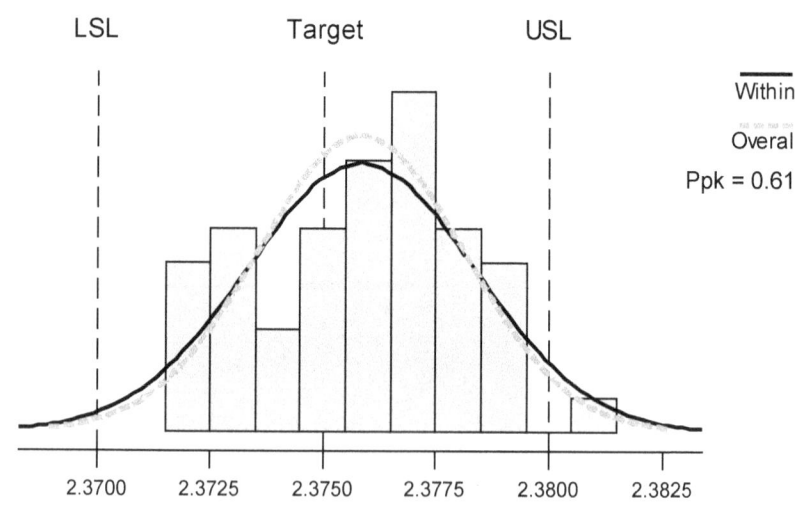

Control Charts: I-MR Chart

This data set plotted with no adjustments .... exhibits in control behavior. Don't make routine adjustments to processes in control. If the process is in control, but is not capable, then the process must be re-engineered to either improve the nominalization and/or reduce normal variation. Use normal statistical tools to perform this: cause & effect diagrams, process mapping, DOEs with standard deviation as a response, COV analysis, etc.

I and MR Chart for F4 w/o

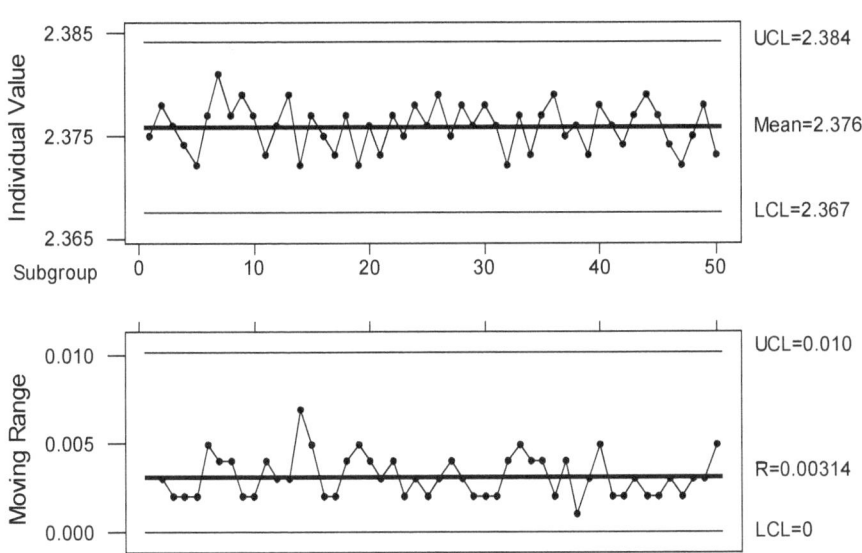

I MR chart above from Minitab®

Control Chart Strategy and Misuse

In the Xbar/R chart below, the range chart represents the "within subgroup" variation. The Xbar chart is between subgroup variation. It is often seen whereby injection molders set up a sub grouping strategy with n=2, 3 or 4 to gather the range chart data. The frequent result however is a Xbar chart that exhibits many points out of control as shown. The Xbar chart calculates it's control limits from the range chart; thus, if you plan subgroups that do not include all the "normal" variation, you get small Rbars and ultra tight X bar control limits (if OOC is too common; charts get ignored). Sometimes there is little or no variation in three consecutive molded shots. The variation caused by resin property drift or heater bands cycling on/off or other plasticizing variation likely takes a longer time than the one minute it takes to mold three shots at today's fast cycle times.

Xbar/R Chart: Y1

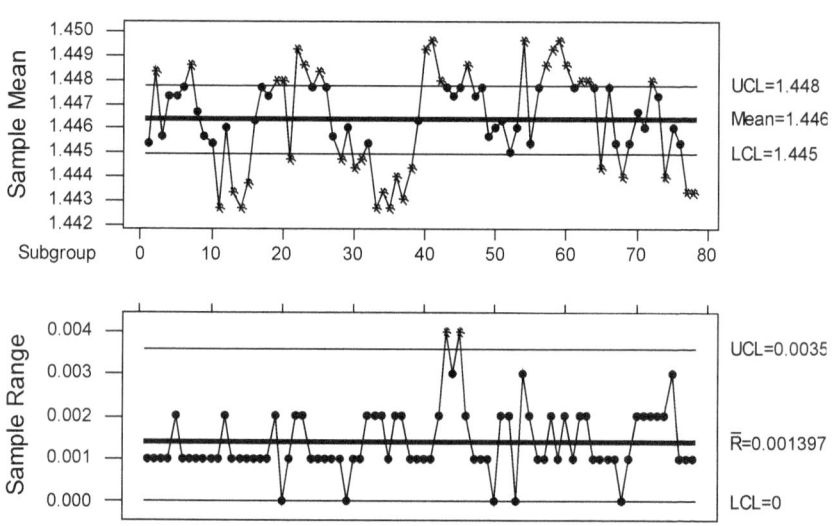

IMPORTANT POINT:

The chart above (Minitab®) has control limits that are 0.003 inches wide ... this writer has seen manufacturing plants whereby a flawed sub-grouping strategy was performed resulting in control limits only 120 millionths wide (0.00012 inches). The aforementioned was actually done deliberately because it results in higher Cpk values which were reported to the customer. If you have a sub-grouping strategy such as n=2 or n=3 (consecutive shots) resulting in very little variation, the estimated sigma will be very small. The calculation for Cpk (not so for Ppk) uses estimated sigma as a divisor resulting in artificially high Cpk values ... AND control limits that are extremely tight since the same falsely low estimated sigma also affects Xbar control limits. The process is nearly always out of control, but the Cpks are much higher than the Ppks! The control chart is worthless for control, but yields great Cpk values to provide a false sense of accomplishment!

Control Chart Strategy: Key Point Revisited

A traditional X-bar & R chart with sub grouping strategy of n=2 or n=3 <u>consecutive shots</u> may be flawed for injection molding because the sub-groups do not meet the definition of rational sub groups ... they are not independent because the barrel holds 3-5 or more shots and mixes that resin (and barrel temperature fluctuations), and this resin (or melt temperature variation) can be a major source of variation in injection molding ... an I-MR chart is preferred!

The moving range chart below has the exact same data as previous page, but uses the difference between sub-groups gathered four hours apart to calculate control limits (3 way chart used only because data already collected that way as n=3; if I-MR chart is appropriate, then use n=1). <u>Some process shift and drift is normal in injection molding; thus, we are letting some of this normal variation into the range chart.</u> Unfortunately, some molders maintain these charts more for developing monthly Cpk values than for control (see next page for Cpk vs Ppk).

I-MR-R/S (Between/Within) Chart: Y1

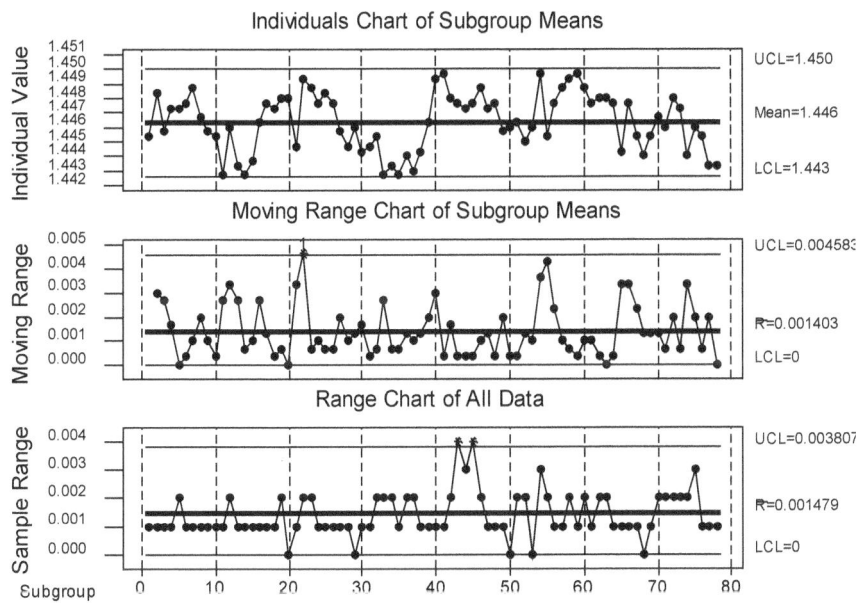

Control Chart Key Points to Remember

1. The Range chart typically drives the control limits for the X bar chart.
2. The estimated std dev may not indicate the actual long term variation.
3. Sub-groups need to be independent relative to sources of variation.
4. Cpk may be very different than Ppk ... Ppk is better overall indicator.
5. Do not compute control limits from data with special causes of variation.
6. Resolve measurement issues before computing control limits.
7. Observe the process and do not over adjust based on first OOC point.
8. Perform root cause analysis for OOC points.
9. Control limit violations do not mean defective product (assumes control limits are inside specification limits). Do not use specification limits as control limits.
10. I-MR charts may be a better sub-grouping strategy for injection molding.

Cpk vs Ppk and Process Drift

The charts seen on previous two pages represent one month of production. As can be seen, there is some drift over the 30 days production. The lack of variation in three consecutive shots creates a small Rbar and subsequently a small estimated std dev (Rbar/d$_2$). This smaller est std dev helps compute a larger Cpk value since the divisor gets smaller. In reality, a Ppk value is much more accurate in showing the total variation and effect on ppm defective: see capability analysis shown below. The Cpk is 2.43, but the Ppk is only 1.02.

Process Capability Analysis for Y1

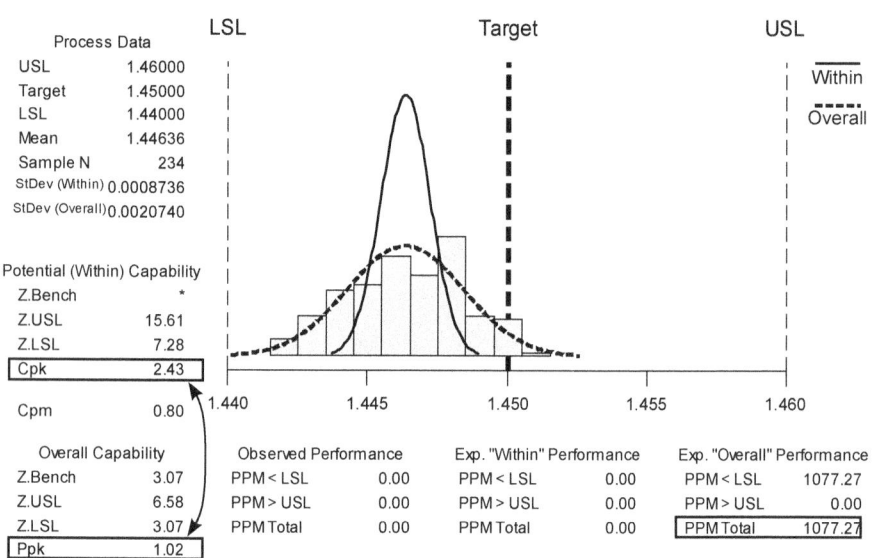

Process Data	
USL	1.46000
Target	1.45000
LSL	1.44000
Mean	1.44636
Sample N	234
StDev (Within)	0.0008736
StDev (Overall)	0.0020740

Potential (Within) Capability	
Z.Bench	*
Z.USL	15.61
Z.LSL	7.28
Cpk	2.43
Cpm	0.80

Overall Capability	
Z.Bench	3.07
Z.USL	6.58
Z.LSL	3.07
Ppk	1.02

Observed Performance		Exp. "Within" Performance		Exp. "Overall" Performance	
PPM < LSL	0.00	PPM < LSL	0.00	PPM < LSL	1077.27
PPM > USL	0.00	PPM > USL	0.00	PPM > USL	0.00
PPM Total	0.00	PPM Total	0.00	PPM Total	1077.27

The Ppk calculation uses a std dev based on all data; whereas, the Cpk calculation assumes the variation is properly represented by Rbar/d$_2$ ... which is based only on variation seen in three consecutive shots or only the within subgroup variation. It is suggested that Ppk be used to best understand the process because it includes within and between subgroup variation – all variation. This is especially important when the data exhibits shift or drift.

Control limits for X bar and R charts frequently use Xbar ± A$_2$ x Rbar, but some software may use ± 3 sigma (estimated or pooled); thus, check your software to fully understand how limits are calculated.

$$\text{Control limits} = \overline{\overline{X}} \pm 3\left(\frac{\overline{R}}{d_2\sqrt{n}}\right)$$

$$\left(\frac{3}{d_2\sqrt{n}}\right) \text{ reduces to } A_2 \text{ fo a given value of n}$$

$$\text{Control limits} = \overline{\overline{X}} \pm A_2\,\overline{R}$$

The above limits are tighter than ± 3 sigma limits as A_2 is 1.880 for n=2 and A_2 gets smaller for higher n values: 1.023 for n=3; compare 1.023 multiplier vs 2.659 seen below:

Control limits for X bar and moving range charts (I-MR):

$$\text{Control limits} = \overline{\overline{X}} \pm E_2 \overline{MR}$$

$$\text{where } E_2 = 2.659$$

$$\frac{3}{d_2} = \frac{3}{1.128} = 2.659 \ldots (\text{for } n = 2 \text{ the } d_2 \text{ is } 1.128)$$

This becomes ± 3 sigma control limits (using est sigma: Rbar/ d_2).

Process Drift and Cycles

What causes data to drift? The graphics below include examples of processes that exhibit such shifts and drifts. Sometimes we never know what causes such variation, but theories include the following:
- Dryer desiccant bed absorption decay then shift to fresh bed
- Hopper loader level variation
- Barrel heater band cycling on/off and/or feed throat cooling
- regrind or virgin resin variation from above or other causes

As can be seen, in most instances, the viscosity changes which results in cavity pressure variation. The effective viscosity is determined by the injection integral (area under plastic pressure fill curve) ... this will be discussed in greater detail in later pages describing real time SPC vs SQC, using pressure transducers and other process monitoring instrumentation. This type equipment (RJG eDart from RJG, Inc, Traverse City, MI) was used to better understand processes shown below.

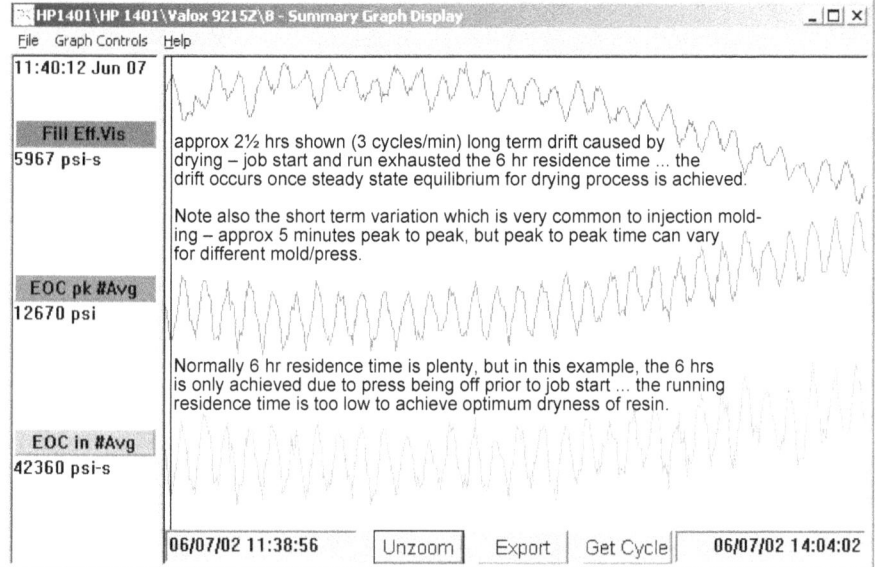

In both of these process summary graphics (above and next page), the cavity pressure does vary as the resin viscosity varies. Pressure transducers located in the cavity measure the cavity pressure. Cavity pressure is a main effect on part sizing due to it's effect on part shrinkage.

Process Drift and Cycles (continued)

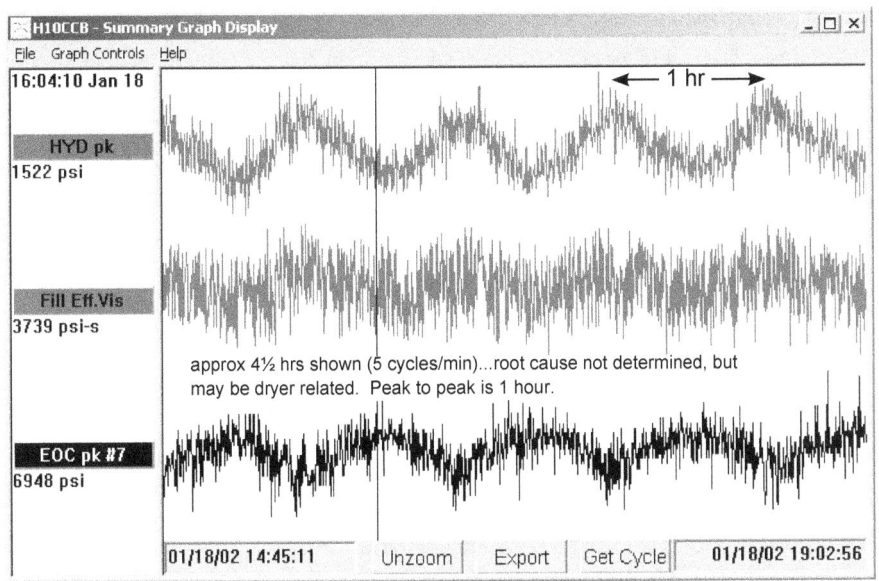

In the graph above, the process looks to exhibit excessive cycling with peaks each hour. This is likely caused by very high thruputs (resin lbs/hr processed) combined with a drier which changes desiccant beds each hour (one hour regeneration time). The solution would be to install a larger hopper (and possibly larger dryer), to yield a greater dryer residence time so all resin is at optimal moisture level during molding.

The dryer performance and engineering should not be under estimated as a source of variation in injection molding of hygroscopic resins such as PET, PC, PMMA, PA, etc.

In addition to dryer engineering, the use of process monitoring equipment is also valuable ... without such, we would not see such variation in order to address it! RJG Inc. is a supplier of such equipment[1].

---

[1] Search RJG on internet as www.rjginc.com ... reference eDart equipment & software. Also contact at: RJG Inc, 3111 Park Drive, Traverse City, MI 49686, 231-947-3111

Injection Molding Variation

Sources of variation can come from the following:

Check ring – Dirt, metal or other FM (foreign material) can get caught in check ring seat area; thus, preventing proper check action which results in resin leakage past check ring – this results in slightly smaller injected shot and reduced packing.

Resin blend variation – Variation in additive package can cause crystallization or release differences. Molecular weight average or distribution can cause flow viscosity differences and resulting molding variation. The resin supplier has a process too that includes variation!

Resin handling – Change in drying, color blending, regrind blending, contamination, etc causing change to melt viscosity. A drying hopper runs low, but refilled by material handler prior to being empty.

Gate blockage – Dirt, metal or other FM may block a small diameter gate; when this happens other cavities may get overpacked because the same plasticized shot gets injected into one (or more) less cavities.

Hyd variation – Flow control and/or pressure valves might be corrupted for a single shot by dirt in valve.

Mold cooling – Molds with small bubblers can have the bubbler become plugged with debris resulting in poor cooling which affects balance of fill, packing and cooling rate. Other cooling circuits may have heat transfer impeded by scale or rust – long term variation.

Mold cooling – Variation in supply coolant temperature to mold and/or back pressure for return lines might affect actual mold temperature. Actual mold temperature affects cooling rate, packing pressure drop and gate freeze time.

Screw wear – Can result in higher melt temperature as plasticizing becomes less efficient resulting in more shear heating; this will also show up as long term variation.

Electrical spike/noise – May corrupt signal and control for a given shot.

Heater band failure – Machine will likely keep running, but heating and plasticizing characteristics may change depending on which heater and proximity to the thermocouple.

Barrel Temperature – The resin temps can vary as heaters cycle on and off ... resin temperatures affect viscosity.

Dirty mold vents – The venting characteristic in a mold may be slowly but constantly changing as the vents get increasingly loaded with volatiles prior to a mold PM.

Ambient conditions – May affect relative humidity which may affect dryer effectiveness; temperature may affect various aspects of machine performance including clamp tonnage, oil temperature, nozzle and barrel temperature. Note also the post mold cooling dynamics affecting shrinkage, sink or void developments.

Imbalanced fill – The thermal homogeneity and flow division at branches affects balance of fill and resulting packing, cooling rates and molded-stress.

Closed loop equipment may be partially successful in minimizing the effects of many of these conditions. Closed loop equipment may vary one parameter to achieve consistency with another. This results in reduced variation for one thing by adding variation to something having less or no detrimental affect (e.g. to achieve the same fill time the fill pressure might be adjusted significantly).

Variation from Cycle Interrupts

The previous page lists many commons sources of molding variation, but it does not list one of the main culprits in causing significant special cause variation – molding cycle interrupts. These interrupts are <u>not</u> common caused normal variation (even though may be common ... meaning frequent ... in some molding plants). The process equilibrium is corrupted in two ways:

1. Longer residence time effects viscosity ... usually downward or easier flowing. Lesser viscosity is especially typical in the case of thermal sensitive resins like polycarbonate, PET, nylon, acetal, acrylic, etc. This lower viscosity may cause the mold to flash.

2. Mold gets colder because residual heat in the mold is removed. Typically a mold is cooled by a coolant such as water; during normal operation, the molding surfaces have some residual heat that has not been removed completely before next cycle starts. After an interrupt, there is often ample time for the coolant to remove all this residual heat; thus, getting the molding surfaces down to the actual coolant temperature. This colder mold temperature often causes a short shot.

As can be seen, the predicted effects from #s 1 & 2 above are conflicting. It all depends on how thermal sensitive the resin is versus how fast the mold is cycling. If mold cycles fast enough there could be considerable residual heat left behind. The net effect is hard to predict, but the equilibrium is certainly compromised.

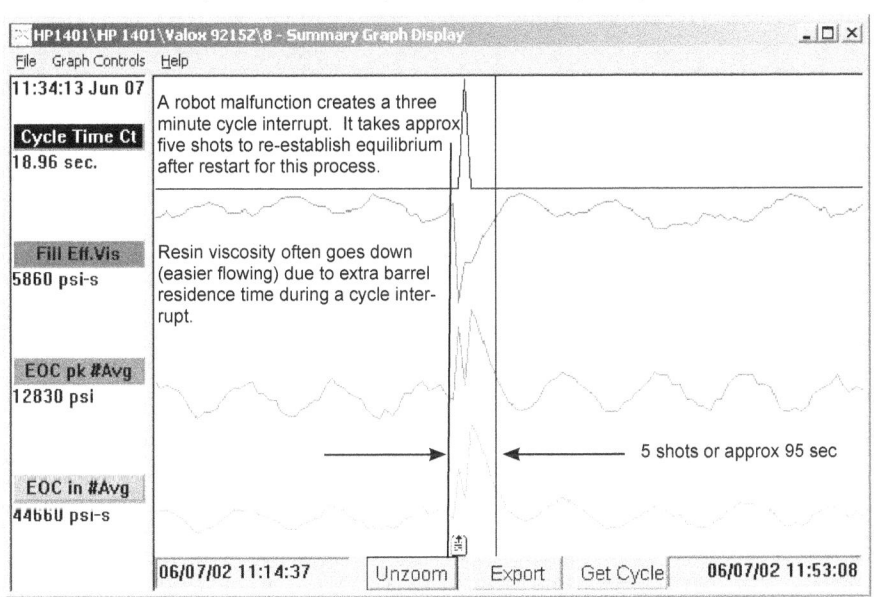

Managing Cycle Interrupts

Most good molders today running critical quality type parts have some method to reject the first few start-up shots after an interrupt (these methods are critical to catching significant molded part variation). These methods include the following:

1. Inject a few airshots to remove low viscosity resin – does not help the colder mold however!
2. Turn off conveyor or robot and manually catch x number of parts.
3. Have press enabled signals that tell flip/flop chute, reversing conveyor or robot to deposit "x" number of shots to an alternate location (scrap bin). Experimentation is needed to determine the "x" number. This "x" quantity is typically between two and five shots depending on percentage shot size used, part features, tolerances, end use requirements and resin type thermal stability.
4. Have optional or auxiliary instrumentation that monitors process (cushion, viscosity, cavity pressure, etc) to direct conveyor or robot to scrap discrepant product (use this in conjunction with the start-up shot rejects).

Obviously, we should also attempt to minimize the cycle interrupts with the following:

1. Properly set mold protection so this feature does not create mold close faults.
2. Set up material loaders and resin conveying system so press does not routinely run out of material. Material supply should be consistent in quality as well – clean, dry, w/o contamination, etc.
3. Make sure parts are removed properly and taken away properly ... even if ejected and dropped, robot unloaded, hand unloaded, etc. Use air blast as needed, multiple eject strokes as needed ... whatever it takes to keep cycle consistent.
4. Use proper discretion when scheduling routine parting line maintenance such as cleaning vents, mold deposits, etc.

Effective SQC

In order to accomplish an effective SQC program, the voice of the process does need to be identified. Often times the product engineer identifies approx three (or more) dimensions to be checked in process. These dimensions are often the longest length, width and height or largest dimensions in the X, Y and Z axis directions for the part.

Later we will discuss DOEs (design of experiments) as used to solve problems, but another use of DOEs is to identify the mold's sensitivity to process variation. The feature exhibiting the most variation during a DOE is likely a good voice of the process with regards to SQC. SQC is statistical quality control and does use statistics and control charts to establish normal variation and distinguish same from special caused variation. SQC typically monitors a part dimension, but is often done after the part is one hour old – this time lag allows for the majority of shrinkage to take place. SPC is statistical process control and is often closer to real time in nature because it's focus is the process rather than the process output – molded parts. We will discuss SPC in more detail later, but since many molders use SQC (even if they call it SPC), we will outline optimum choices for SQC.

Effective SQC ... DOE Data Example 1

<table>
<tr><td colspan="16" align="center">DOE Data Example #1</td></tr>
<tr><td></td><td></td><td></td><td></td><td>Peak</td><td>Core</td><td>Cyc</td><td>C7</td><td>C7</td><td colspan="3">control dimensions</td><td>Visual</td><td></td><td></td></tr>
<tr><td>SO</td><td>RO</td><td>CP</td><td>BLK</td><td>C7 psi</td><td>Temp</td><td>Time</td><td>Cheek flat</td><td>Flatness</td><td>L - avg</td><td>H - avg</td><td>W - avg</td><td>1-5</td><td>C7_C.I.</td><td>C7_Peak</td></tr>
<tr><td>1</td><td>8</td><td>1</td><td>1</td><td>9720</td><td>60</td><td>18</td><td>0.0328</td><td>0.0509</td><td>44.254</td><td>37.693</td><td>25.513</td><td>2.0</td><td>30210</td><td>9600</td></tr>
<tr><td>2</td><td>9</td><td>1</td><td>1</td><td>12720</td><td>60</td><td>18</td><td>0.0227</td><td>0.0346</td><td>44.254</td><td>37.702</td><td>25.528</td><td>2.5</td><td>43599</td><td>12586</td></tr>
<tr><td>3</td><td>4</td><td>1</td><td>1</td><td>9720</td><td>80</td><td>18</td><td>0.0310</td><td>0.0626</td><td>44.258</td><td>37.748</td><td>25.527</td><td>1.9</td><td>46077</td><td>9603</td></tr>
<tr><td>4</td><td>5</td><td>1</td><td>1</td><td>12720</td><td>80</td><td>18</td><td>0.0227</td><td>0.0505</td><td>44.257</td><td>37.776</td><td>25.520</td><td>2.6</td><td>44310</td><td>12793</td></tr>
<tr><td>5</td><td>2</td><td>1</td><td>1</td><td>9720</td><td>60</td><td>22</td><td>0.0437</td><td>0.0288</td><td>44.261</td><td>37.770</td><td>25.517</td><td>3.5</td><td>30721</td><td>9687</td></tr>
<tr><td>6</td><td>3</td><td>1</td><td>1</td><td>12720</td><td>60</td><td>22</td><td>0.0250</td><td>0.0227</td><td>44.262</td><td>37.803</td><td>25.517</td><td>4.1</td><td>45873</td><td>12784</td></tr>
<tr><td>7</td><td>6</td><td>1</td><td>1</td><td>9720</td><td>80</td><td>22</td><td>0.0389</td><td>0.0338</td><td>44.265</td><td>37.759</td><td>25.527</td><td>3.1</td><td>30829</td><td>9689</td></tr>
<tr><td>8</td><td>7</td><td>1</td><td>1</td><td>12720</td><td>80</td><td>22</td><td>0.0248</td><td>0.0243</td><td>44.264</td><td>37.791</td><td>25.521</td><td>4.4</td><td>46077</td><td>12939</td></tr>
<tr><td>9</td><td>1</td><td>0</td><td>1</td><td>11220</td><td>70</td><td>20</td><td>0.0315</td><td>0.0335</td><td>44.260</td><td>37.774</td><td>25.522</td><td>3.4</td><td>36653</td><td>11236</td></tr>
<tr><td>10</td><td>12</td><td>1</td><td>2</td><td>9720</td><td>60</td><td>18</td><td>0.0354</td><td>0.0583</td><td>44.257</td><td>37.749</td><td>25.516</td><td>2.0</td><td>28764</td><td>9706</td></tr>
<tr><td>11</td><td>10</td><td>1</td><td>2</td><td>12720</td><td>60</td><td>18</td><td>0.0232</td><td>0.0367</td><td>44.256</td><td>37.787</td><td>25.525</td><td>3.0</td><td>43630</td><td>12659</td></tr>
<tr><td>12</td><td>13</td><td>1</td><td>2</td><td>9720</td><td>80</td><td>18</td><td>0.0367</td><td>0.0663</td><td>44.257</td><td>37.741</td><td>25.530</td><td>1.5</td><td>29218</td><td>9816</td></tr>
<tr><td>13</td><td>17</td><td>1</td><td>2</td><td>12720</td><td>80</td><td>18</td><td>0.0252</td><td>0.0503</td><td>44.260</td><td>37.776</td><td>25.527</td><td>2.3</td><td>42976</td><td>12738</td></tr>
<tr><td>14</td><td>16</td><td>1</td><td>2</td><td>9720</td><td>60</td><td>22</td><td>0.0476</td><td>0.0327</td><td>44.262</td><td>37.761</td><td>25.522</td><td>2.9</td><td>29996</td><td>9777</td></tr>
<tr><td>15</td><td>14</td><td>1</td><td>2</td><td>12720</td><td>60</td><td>22</td><td>0.0278</td><td>0.0223</td><td>44.261</td><td>37.802</td><td>25.531</td><td>3.9</td><td>44962</td><td>12603</td></tr>
<tr><td>16</td><td>18</td><td>1</td><td>2</td><td>9720</td><td>80</td><td>22</td><td>0.0456</td><td>0.0342</td><td>44.263</td><td>37.765</td><td>25.523</td><td>2.9</td><td>29897</td><td>9814</td></tr>
<tr><td>17</td><td>11</td><td>1</td><td>2</td><td>12720</td><td>80</td><td>22</td><td>0.0270</td><td>0.0245</td><td>44.261</td><td>37.802</td><td>25.531</td><td>3.7</td><td>44573</td><td>12776</td></tr>
<tr><td>18</td><td>15</td><td>0</td><td>2</td><td>11220</td><td>70</td><td>20</td><td>0.0339</td><td>0.0367</td><td>44.263</td><td>37.765</td><td>25.523</td><td>2.5</td><td>35118</td><td>11165</td></tr>
<tr><td>correl</td><td></td><td></td><td></td><td></td><td></td><td></td><td>-0.85</td><td>-0.28</td><td>-0.05</td><td>0.46</td><td>0.34</td><td>0.42</td><td>1.00</td><td></td></tr>
<tr><td>correl</td><td></td><td></td><td></td><td></td><td></td><td></td><td>-0.84</td><td>-0.45</td><td>-0.01</td><td>0.52</td><td>0.27</td><td>0.51</td><td></td><td>1.00</td></tr>
<tr><td>average</td><td></td><td></td><td></td><td></td><td></td><td></td><td></td><td></td><td>44.260</td><td>37.765</td><td>25.523</td><td></td><td></td><td></td></tr>
<tr><td>nominal</td><td></td><td></td><td></td><td></td><td></td><td></td><td></td><td></td><td>44.250</td><td>37.750</td><td>25.578</td><td></td><td></td><td></td></tr>
<tr><td colspan="9">desirability (aka nominalization or process centering)</td><td>93.5%</td><td>90.2%</td><td>56.3%</td><td></td><td></td><td></td></tr>
<tr><td>max</td><td></td><td></td><td></td><td></td><td></td><td></td><td></td><td></td><td>44.265</td><td>37.803</td><td>25.531</td><td></td><td></td><td></td></tr>
<tr><td>min</td><td></td><td></td><td></td><td></td><td></td><td></td><td></td><td></td><td>44.254</td><td>37.693</td><td>25.513</td><td></td><td></td><td></td></tr>
<tr><td>range</td><td></td><td></td><td></td><td></td><td></td><td></td><td></td><td></td><td>0.011</td><td>0.110</td><td>0.018</td><td></td><td></td><td></td></tr>
<tr><td>USL</td><td></td><td></td><td></td><td></td><td></td><td></td><td></td><td></td><td>44.400</td><td>37.900</td><td>25.703</td><td></td><td></td><td></td></tr>
<tr><td>LSL</td><td></td><td></td><td></td><td></td><td></td><td></td><td></td><td></td><td>44.100</td><td>37.600</td><td>25.453</td><td></td><td></td><td></td></tr>
<tr><td>TOL</td><td></td><td></td><td></td><td></td><td></td><td></td><td></td><td></td><td>0.300</td><td>0.300</td><td>0.250</td><td></td><td></td><td></td></tr>
<tr><td>range/tol</td><td></td><td></td><td></td><td></td><td></td><td></td><td></td><td></td><td>3.7%</td><td>36.7%</td><td>7.2%</td><td></td><td></td><td></td></tr>
</table>

As mentioned above, we must identify the voice of the process in terms of measurable part features whether it be part weight, length, height, diameter, etc. A DOE which introduces planned variation into the experiment can be useful to identify the best dimensional voice of the process. If DOEs are completely new to you, see DOE pages later in this book. In the DOE data above, it can be seen by looking at the range and range ÷ tolerance that the height (H-avg) is the best voice of the process (of the three dimensions compared).

Also shown in this table is the nominalization. There is excellent steel nominalization for L-avg & H-avg, this part is at a low risk for defects, but the width could be improved, but if variation is low enough this centering might still yield outstanding Ppks and Z scores. This data set is from a DOE with considerable planned variation; thus, we did not compute Z scores.

The height check (H-avg) would be the best choice for SQC to determine when special cause variation is attempting to corrupt the quality of the parts. We choose this based on the range/tol (dark shaded cells) because the H-avg range/tol equals 37% and is much more sensitive to process variation than other dimensions. If centering is > 75% and corresponding range/tol < 25%, then the feature is not a good voice of the process and/or is not a high risk; thus, monitor for 3 months and drop from routine Q.C. checks.

Effective SPC

In this example, a pressure transducer was placed to read pressure inside cavity #1. As we calculate the data for nominalization, range used, means, etc ... we also can compute the correlation between the recorded peak cavity pressure and cycle integral – area under the curve (see graphic next page). There is a good correlation between CI (cycle integral) and peak vs length and height, but not width. It should be noted here that correlation is not necessarily a cause and effect relationship, but may be ... in this case it is, and it often is in injection molding. See later pages on regression analysis to develop predictive equations. Since this correlation does exist, we can monitor peak cav psi or cycle integral as a real time SPC monitoring method. This will save inspection time and provide quicker feedback. We can perform the afore-mentioned regression analysis to develop a predictive math equation as to where the limits should be set to stay within specification limits.

DOE Data Example #2										
				Pack	Mold	Fill	control dimensions			
SO	RO	CP	BLK	psi	Temp	Time	L - avg	H - avg	W - avg	C1_C.I.	C1_Peak
1	8	1	1	5500	100	0.60	4.2148	0.8722	1.5015	12416	4612
2	9	1	1	7000	100	0.60	4.2235	0.8732	1.5018	15612	5611
3	4	1	1	5500	120	0.60	4.2155	0.8725	1.5016	12878	4684
4	5	1	1	7000	120	0.60	4.2276	0.8750	1.5015	16245	5916
5	2	1	1	5500	100	1.00	4.2142	0.8711	1.5016	12156	4580
6	3	1	1	7000	100	1.00	4.2189	0.8729	1.5016	15410	5404
7	6	1	1	5500	120	1.00	4.2198	0.8725	1.5016	13210	4788
8	7	1	1	7000	120	1.00	4.2236	0.8741	1.5018	15890	5712
9	1	0	1	6250	110	0.80	4.2219	0.8735	1.5017	14112	5068
correl							0.89	0.86	0.41	1.00	
correl							0.91	0.87	0.37		1.00
average							4.220	0.873	1.502		
nominal							4.225	0.875	1.500		
desirability (aka nominalization or process centering)							-0.4%	60.0%	67.3%		
max							4.228	0.875	1.502		
min							4.214	0.871	1.502		
range							0.013	0.004	0.000		
USL							4.230	0.880	1.505		
LSL							4.220	0.870	1.495		
TOL							0.010	0.010	0.010		
range/tol							134.0%	39.0%	3.0%		

Effective SPC .. Cycle Integrals

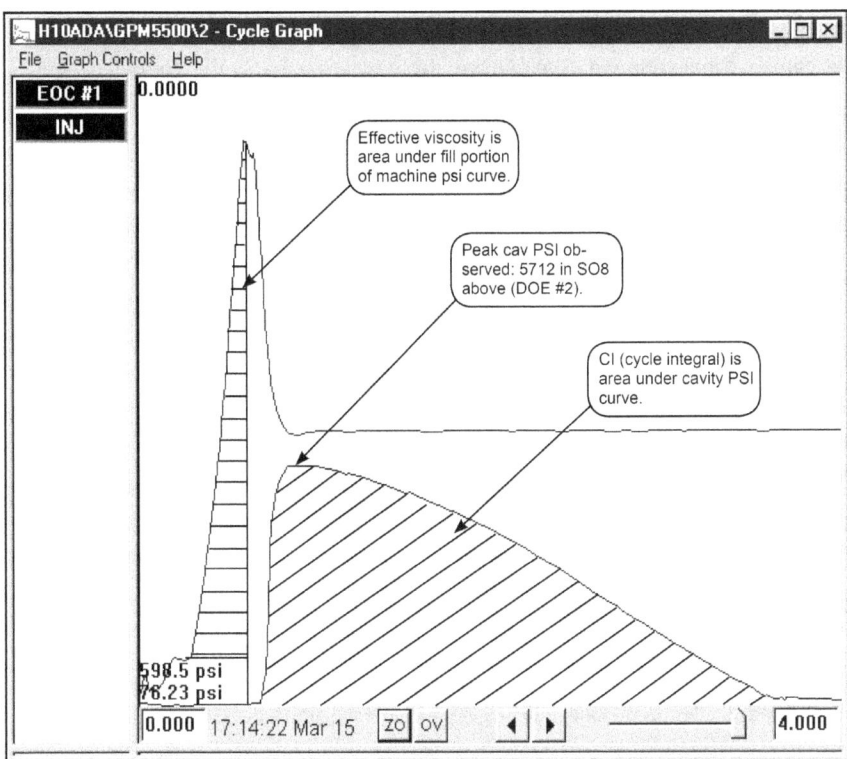

The data provided in graph above includes considerable data which well describes or "fingerprints" the process. The horizontal bars cross hatched area is the area under fill portion of machine's injection hydraulic pressure. The area is computed by math integration and is affected by pressure, speed, temperature, resin molecular weight, resin drying, VP transfer, etc; thus, a comprehensive process data output to monitor ... and is real time! Real time means we know before the mold opens if the part is good or not (or at least high probability of being good).

The diagonal cross hatched area indicates the in-the-cavity pressure (5,712 peak psi) inside the molded part wall with decay over time as part cools ... and also indicates the cycle Integral – area under cavity pressure curve. This is also real time and comprehensive. RJG process monitoring and control instrumentation can be used for real time SPC, but does require a higher skill level by personnel and a few process experiments such as DOE to characterize the process and part's sensitivity to process variation. The CI data (cycle integral data) was shown on previous page to correlate well with part dimensions, but advanced skills are needed to perform these DOE studies and the data analysis required.

Real Time SPC

The goal for SPC is to be real time. A process in control permits the future to be predicted. A process that contains only normal variation should continue to exhibit that same normal variation. We can use this concept to study the normal, in control process, and set limits for the future; be sure to output the alarm violations to take action such as diverter chutes or robot control.

Later in the book, it will be shown that 6σ short term may only be 4.5σ long term. This long term data shift and drift has been predicted via calculation to be 1.5σ. If we then apply $\pm 4.5\sigma$ (3 +1.5 allowance for shift) limits to the process output being monitored, we should have a good basis for SPC control. Violations will signal special cause variation needing root cause analysis. Note: Resin lot changes can be the special cause; thus, be prepared to review and change process limits as needed for effective SPC. Studies can be done with regression analysis and/or DOEs to develop predictive equations whereby the required pack psi or melt temp can be changed to compensate for resin variation.

Real Time SPC Monitoring (in approximate order of preference)
(result from misc process inputs or combination thereof)
1.   CI (cycle integral based on actual cav psi); <u>requires transducer</u>. This is a very comprehensive value in that it is strongly affected by peak cav psi, cooling rate, resin viscosity, pack time, etc and many of the inputs driving the following process outputs.
2.   Injection fill time (very important) – should be consistent unless transfer by cav psi, then depends on balance of fill consistency.
3.   Actual maximum filling pressure; specifically the VPT cutoff psi. Take care that injection start is not the peak; do not confuse your interpretation of this readout; some processes may decelerate enough at end of fill whereby VPT is not the peak pressure)
     • may indicate resin viscosity changes (resin or dryer)
     • may indicate gate blockages
     • may indicate mold temperature problems
     • may indicate other machine problems
4.   Pack and/or hold pressure – should be consistent; if not, indicates machine problem. The pack greatly affects shrinkage and resultant part sizing; thus, this is critical, but <u>should be easy for press to control</u> (should rarely be OOC).
5.   Actual melt temperature (front zone if actual melt temp is not available)
6.   Mold Temperature (actual or coolant ea half).
7.   Cushion
     • may indicate check ring problems or plasticizing problems
     • may indicate self blocked cavities
     • may indicate resin or dryer problem
8.   Actual cutoff position
     • may indicate transducer problem (if transfer is by cav psi)
     • may indicate need for fill speed adjust
9.   Actual shot size -- should be consistent, but if not might indicate a decompression problem which can affect shot size and resulting cushion ... #7,8 & 9 all work together, problem with one can affect the others.
10. Cycle time, hold time, plasticize time, etc ... important, but <u>should be easy to control</u> effectively; plasticize time should be only source of variation; thus, first among these time choices.
     • may indicate plasticizing problem
     • may indicate aux equip problem

Transducer Choices & Real Time SPC

Transducer configuration/construction types:

1. Piezoelectric transducers have a quartz crystal which produces a charge as it's face receives a "load". Kistler transducers are piezoelectric type transducers. The charge produced is small; thus, a charge amplifier is required.
2. Strain gage transducers, which include a wheatstone bridge, which takes an input voltage of 10 VDC, and delivers an altered output based on the loading; once calibrated properly the voltage output correlates to a pressure in cavity.

Transducer installation/application types

The "load" can be a force or a pressure:

1. Force type applications – these are installations where a transducer is under an ejector pin. The cavity pressure acts on the face of the selected ejector pin which creates a force acting on the transducer. The area of ejector pin (in²) cancels out the in² in lbs/in²; thus resulting in only a force of lbs acting on the transducer. In these applications, we must convert the force back to pressure somewhere in the setup by listing the pin diameter so machine or instrumentation software can convert the force signal back to pressure signal. Remember: force is lbs (or Newtons) and pressure is lbs/in² (or bar).

2. Pressure type applications – when there is a direct read transducer whereby the transducer face protrudes thru the core or cavity surface, it is a pressure transducer.

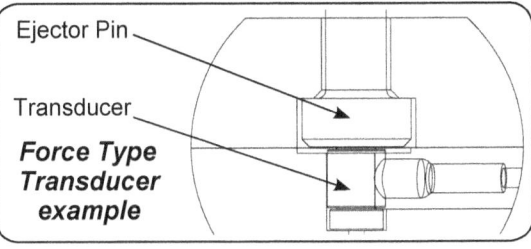

Ejector Pin

Transducer

Force Type Transducer example

These transducers directly see the cavity pressure; thus, there is no need to enter a diameter to convert a force to a pressure.

Transducer locations:

1. For controlling the VP transfer, the transducer should be located closer to the gate, but after it (aka "post gate"). This is because if located at end of fill, there may not be adequate response time to perform transfer before the cavity sees a quick rise in cavity pressure. During fast, or even moderate fill times, there is some time before the injection ram and screw slow down. This is caused by inertia in the mass moving forward AND the electrical and mechanical response time of signals and hydraulic valves to fully

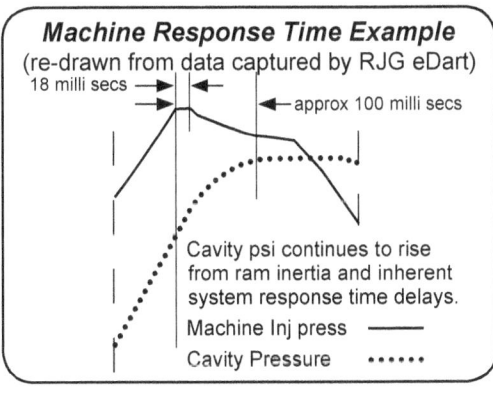

Machine Response Time Example
(re-drawn from data captured by RJG eDart)

18 milli secs →

←— approx 100 milli secs

Cavity psi continues to rise from ram inertia and inherent system response time delays.

Machine Inj press ——
Cavity Pressure ••••••

implement the change. If the transducer was at end of fill, we might overpack the cavity in these cases due to lack of excess volume to absorb the pressure.

2. For monitoring part quality (real time SPC), the transducer should be located closer to end of fill. This better monitors the effects of packing pressure and resulting curve decay which is affected by packing and cooling rates.

Blank Page

Attribute Control Charts

All control charts – including attribute charts – are designed to achieve process control and improvement by supplying timely feedback as to the current state of process performance. The key to success is to perform root cause analysis, and then implement corrective actions.

Attribute control charts are used for non variable type data. This non variable or non numerical data is attribute data such as good vs bad ... there is no numerical scale describing how good or how bad, just good vs bad.

There are four types of control charts to choose from for attributes:

1. p charts – plots percent defective; can be used when subgroups are not equal sized since it is percentage.

2. np charts – plots actual numbers of scrapped parts (not %), but must have subgroups of equal size; since we are plotting the actual numbers of scrapped parts (instead of percent), the results may be easier to interpret or relate to.

3. u charts – plots the average number of defects per subgroup (can be used with different subgroup sizes)

4. c charts – plots the count or number of defects per unit in a subgroup (subgroups must stay the same size) ... different than np which is counting scrapped parts; with c charts we might be counting the number of black specks per part or per hours worth of parts or three parts taken at random within an hour, etc. With this approach, we might get better insight as to the severity of the defect type, but requires more time to count and track. A scrapped part in the p or np chart would have a value of one part scrapped, but with the c chart we could capture how bad is bad – such as 7 black specks vs 1 black speck (maybe the threshold for a scrap part is one single black speck).

P - Charts

A p chart is a control chart for attributes. The p indicates that the "p" or percentage defects is being charted. Since the p charts convert numbers of defectives to percent defective, there can be different sample sizes present in the data.

The percent defects is simply the number of defects divided by the sample size, such as parts per hour.

$$p = \text{percent defective} = \frac{\text{\# defects}}{\text{sample size}} = \frac{x}{n}$$

We can also calculate a p-bar as follows:

$$\text{p-bar} = \bar{p} = \frac{\sum p_i}{k} \quad (k = \text{\# of subgroups or samples})$$

or

$$\bar{p} = \frac{\text{sum of defects in all samples (x)}}{\text{sum of sample sizes (n)}} = \frac{\sum x}{\sum n}$$

We can establish control limits based on ± 3 sigma limits. The formula for p-sigma is as follows:

$$\text{p-sigma} = \sigma_p = \sqrt{\frac{\bar{p} \times (1 - \bar{p})}{n}}$$

Note: if the sample size n varies each hour; then compute the average sample size and use for n

Frequently, the lower control limit (LCL) will default to zero. The formulas for UCL and LCL are as follows:

$$\text{p-LCL} = \bar{p} - 3 \cdot \sigma_p$$

$$\text{p-LCL} = \bar{p} - 3 \cdot \sqrt{\frac{\bar{p} \cdot (1 - \bar{p})}{n}}$$

Note: use zero if above is less than zero

$$\text{p-UCL} = \bar{p} + 3 \cdot \sigma_p$$

$$\text{p-UCL} = \bar{p} + 3 \cdot \sqrt{\frac{\bar{p} \cdot (1 - \bar{p})}{n}}$$

See next page for a sample data set, control chart and Minitab + Excel solutions.

P - Charts in Minitab

Graphs made using MINITAB® v.16.1.1 from Minitab Inc.

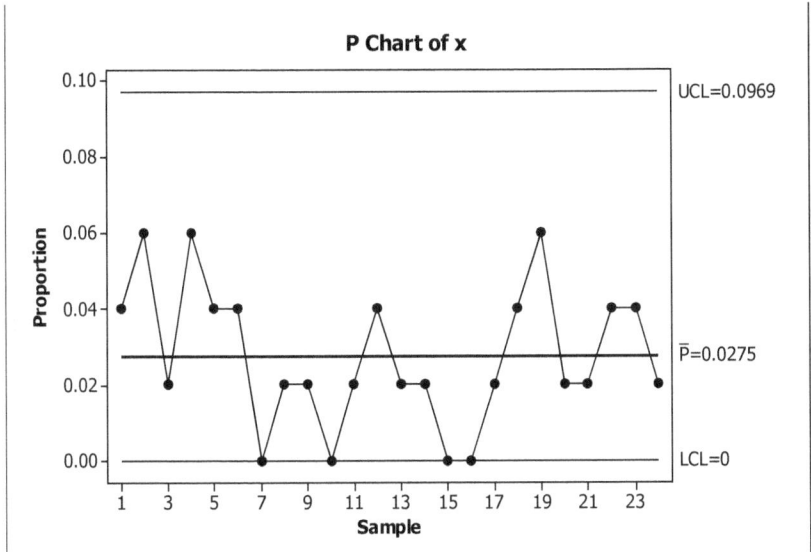

P Chart of x

The defect in column A could be black specks, sinks, etc.

A Minitab graph can be made by simply listing the defects in a column (such as column A at right); then select Stat, Control Charts, Attribute Charts, P... then identify the column with defect data and enter the subgroup size (sample size 50 in this example[1]) ... or identify column w/ sample sizes.

An Excel spreadsheet could be arranged as shown at right ... formulas are listed below:

cell A27=COUNT(A3:A26)

cell B27=AVERAGE(B3:B26)

cell C27 =AVERAGE(C3:C26)

cell A30=SUM(A3:A26)

cell B31=SUM(B3:B26)

cell C33=SQRT((C27*(1-C27))/B27)

cell C34=C27+3*C33

cell C35=IF((C27-3*C33)< 0, 0, (C27-3*C33))

	A	B	C
1	# defects	sample size	calc % - p
2	x	n	p
3	2	50	0.04
4	3	50	0.06
5	1	50	0.02
6	3	50	0.06
7	2	50	0.04
8	2	50	0.04
9	0	50	0.00
10	1	50	0.02
11	1	50	0.02
12	0	50	0.00
13	1	50	0.02
14	2	50	0.04
15	1	50	0.02
16	1	50	0.02
17	0	50	0.00
18	0	50	0.00
19	1	50	0.02
20	2	50	0.04
21	3	50	0.06
22	1	50	0.02
23	1	50	0.02
24	2	50	0.04
25	2	50	0.04
26	1	50	0.02
27	24	50.0	0.02750
28	(k)	(n)	(pbar)
29			
30	33	<-sum (x)	
31		1200	<-sum(n)
32			
33		(sigma-p) ->	0.02313
34		UCL->	0.096882
35		LCL->	0

---

[1] In this example, sample sizes are equal, but do not have to be for p charts

NP - Charts

A np chart is a control chart for attributes. The np chart plots the number of defects and does not require conversion to percent defective. The np chart does require that the sample size be consistently the same size.

The value plotted on the control chart is simply the actual number of defects - not a percentage.

We will calculate a np-bar to be the center line (CL) as follows:

$$np\text{-bar} = n \cdot \bar{p}$$

or

$$\frac{\Sigma x}{k} \quad (k = \text{\# of subgroups})$$

We will still need to compute a p-bar for use in computing the control limits:

$$p\text{-bar} = \bar{p} = \frac{\Sigma x}{n \cdot k} \quad (k = \text{\# of subgroups or samples})$$

or

$$\bar{p} = \frac{\text{sum of all defects in all samples (x)}}{\text{sum of sample sizes (n)}} = \frac{\Sigma x}{\Sigma n}$$

Frequently, the lower control limit (LCL) will default to zero. The formulas for UCL and LCL are as follows:

$$UCL = n \cdot \bar{p} + 3 \cdot \sqrt{n \cdot \bar{p} \cdot (1-\bar{p})}$$

$$LCL = n \cdot \bar{p} - 3 \cdot \sqrt{n \cdot \bar{p} \cdot (1-\bar{p})}$$

Note: use zero if above is less than zero

One might argue that the LCL should never be higher than zero since that should be the ultimate goal - zero defects, but we still compute a LCL to trigger a root cause investigation (even though results are favorable) to identify possible best practices.

See next page for a sample data set, control chart and Minitab + Excel solutions.

NP - Charts in Minitab

Graphs made using MINITAB® v.16.1.1 from Minitab Inc.

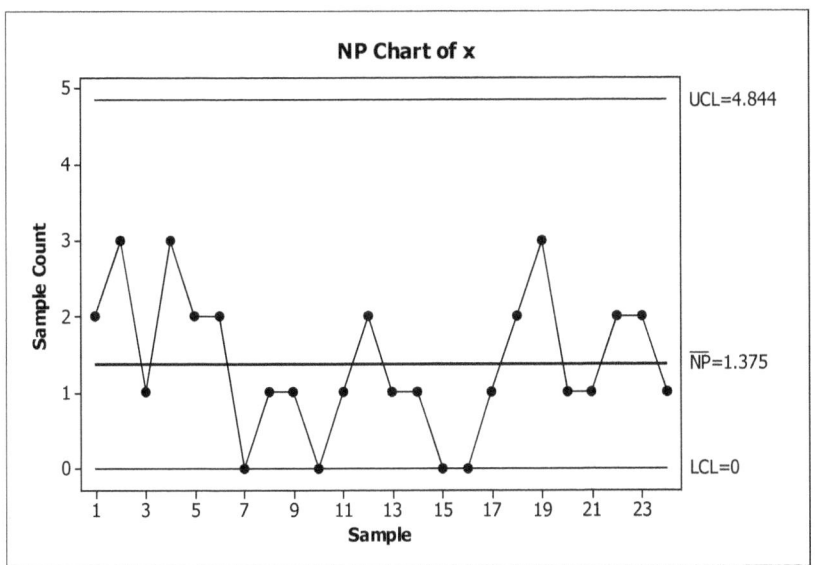

A Minitab graph can be made by simply listing the defects in a column (such as column A at right); then select Stat, Control Charts, Attribute Charts, NP... then identify the column with defect data and enter the column with subgroup size (sample size).

An Excel spreadsheet could be arranged as shown at right ... formulas are listed below:

cell A27=COUNT(A3:A26)

cell B27=AVERAGE(B3:B26)

cell A29=SUM(A3:A26)

cell A30=AVERAGE(A3:A26)

cell A31=A32*B27

cell A32=A29/(A27*B27)

cell A33=B27*A32+3*SQRT((B27*A32*(1-A32)))

	A	B
1	# defects	sample size
2	x	n
3	2	50
4	3	50
5	1	50
6	3	50
7	2	50
8	2	50
9	0	50
10	1	50
11	1	50
12	0	50
13	1	50
14	2	50
15	1	50
16	1	50
17	0	50
18	0	50
19	1	50
20	2	50
21	3	50
22	1	50
23	1	50
24	2	50
25	2	50
26	1	50
27	24	50.0
28	(k)	(n)
29	33	<-sum (x)
30	1.375	<-avg (x)= np-bar
31	1.375	<-avg (x)= np-bar
32	0.0275	<-p-bar
33	4.844	<-UCL
34	0.000	<-LCL

cell A34=IF(B27*A32-3*SQRT((B27*A32*(1-A32)))<0, 0, B27*A32-3*SQRT((B27*A32*(1-A32))))

U - Charts

A u chart can be used to plot the average number of defects per unit. A u chart is used with different subgroup sizes as shown below.

	A	B	C	D	E
1	parts/hr	specks	specks/unit		
2	n	c	u	UCL	LCL
3	143	16	0.112	0.172	0.017
4	144	12	0.083	0.172	0.018
5	144	11	0.076	0.172	0.018
6	142	10	0.070	0.172	0.017
7	140	11	0.079	0.173	0.017
8	144	11	0.076	0.172	0.018
9	78	6	0.077	0.199	0.000
10	140	9	0.064	0.173	0.017
11	145	15	0.103	0.171	0.018
12	143	15	0.105	0.172	0.017
13	143	12	0.084	0.172	0.017
14	143	11	0.077	0.172	0.017
15	144	11	0.076	0.172	0.018
16	144	9	0.063	0.172	0.018
17	145	11	0.076	0.171	0.018
18	16	2	0.125	0.325	0.000
19	96	5	0.052	0.189	0.000
20	144	24	0.167	0.172	0.018
21	145	32	0.221	0.171	0.018
22	143	33	0.231	0.172	0.017
23	143	9	0.063	0.172	0.017
24	144	8	0.056	0.172	0.018
25	144	8	0.056	0.172	0.018
26	144	12	0.083	0.172	0.018
27	n	c			
28	3201	303			
29		0.0947	<<< u-bar		

cell C3=B3/A3

cell D3=$B$29+3*(SQRT(B$29)*(1/SQRT(A3)))

cell E3=IF($B$29-3*(SQRT(B$29)*(1/SQRT(A3)))<0,0,($B$29-3*(SQRT(B$29)*(1/SQRT(A3)))))

cell B29=B28/A28

cells C3, D3 & E3 are formatted such that you can copy down thru row 26 .... see next page for Minitab graph of same data

U - Charts in Minitab

The graph at bottom comes from the data set at right (same as previous Excel data set seen on previous page).

n	Specks (c)
143	16
144	12
144	11
142	10
140	11
144	11
78	6
140	9
145	15
143	15
143	12
143	11
144	11
144	9
145	11
16	2
96	5
144	24
145	32
143	33
143	9
144	8
144	8
144	12

$$\bar{u} = \frac{\text{sum of n's}}{\text{sum of c's}}$$

$$UCL = \bar{u} + \frac{3 \times \sqrt{\bar{u}}}{\sqrt{n}}$$

$$LCL = \bar{u} - \frac{3 \times \sqrt{\bar{u}}}{\sqrt{n}}$$

A Minitab graph can be made by simply listing the defects per unit (specks) in a column (such as column B - previous page and at right) ... list also the subgroup sizes in a column; then select Stat, Control Charts, Attribute Charts, U... then identify the column with defect data (Variables) and enter the column with sub-group size (column A is the sample size).

Graphs made using MINITAB® v.16.1.1 from Minitab Inc.

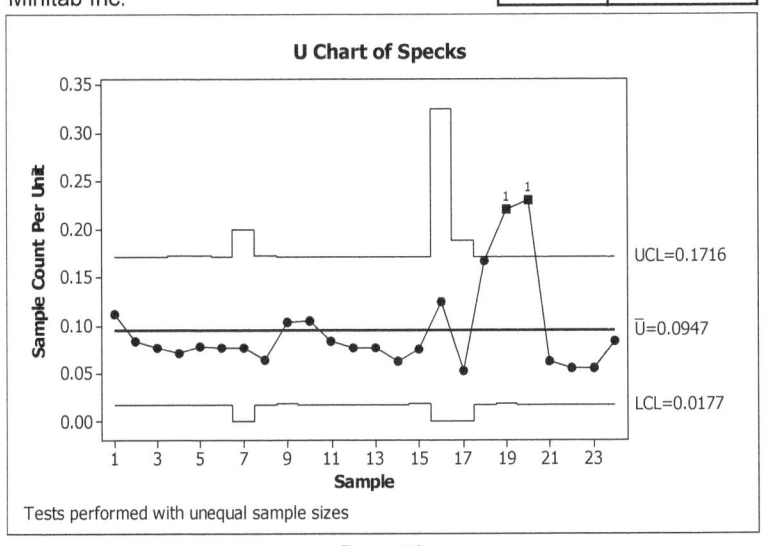

U Chart of Specks

Tests performed with unequal sample sizes

C - Charts

A c chart is a control chart for attributes. The c chart plots the number of defects per unit ... such as per part, per sq foot, per hour, per day, etc.

Since this is defects per hour, we get a better accounting for the overall severity of the defect type. A scrapped part might have 1 black speck causing the part to be scrapped or 5 black specks ... either way one part is scrapped, but it may be good to know what the true void incidence rate is. The downside is the time required to actually count the black specks.

When you use a c-chart, the <u>sample size should be consistent</u>. On next page, the chart is per hour, and each hour represents 70 pcs produced.

We will calculate a c-bar to be the center line (CL) as follows:

$$\text{c-bar} = \bar{c} = \frac{\Sigma c}{k} \text{ (k = \# of subgroups or units)}$$

The c-sigma is computed as follows:

$$\text{c-sigma} = \sigma_c = \sqrt{\bar{c}} = \sqrt{\text{c-bar}}$$

Frequently, the lower control limit (LCL) will default to zero. The formulas for UCL and LCL are as follows:

$$\text{UCL} = \bar{c} + 3 \cdot \sqrt{\bar{c}}$$

$$\text{LCL} = \bar{c} - 3 \cdot \sqrt{\bar{c}}$$

See next page for a sample data set, control chart and Minitab + Excel solutions.

C - Charts in Minitab

Graphs made using MINITAB® v.16.1.1 from Minitab Inc.

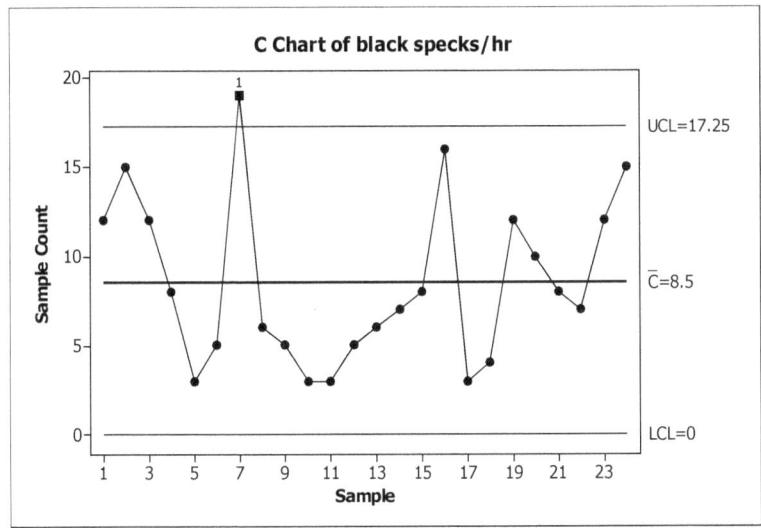

A Minitab graph can be made by simply listing the defects in a column (such as column A at right); then select Stat, Control Charts, Attribute Charts, C... then identify the column with defect data.

An Excel spreadsheet could be arranged as shown at right ... formulas are listed below:

cell A27=COUNT(A3:A26)

cell A29=SUM(A3:A26)

cell A30=AVERAGE(A3:A26)

cell A31=A29/A27

cell A32=SQRT(A30)

cell A33=A30+3*A32

cell A34=IF((A30-3*A32)<0, 0, (A30-3*A32))

	A	B
1	black specks/hr	
2	c	
3	12	
4	15	
5	12	
6	8	
7	3	
8	5	
9	19	
10	6	
11	5	
12	3	
13	3	
14	5	
15	6	
16	7	
17	8	
18	16	
19	3	
20	4	
21	12	
22	10	
23	8	
24	7	
25	12	
26	15	
27	24	
28	(k)	
29	204	<-sum (c)
30	8.5	<-avg (c)= c-bar
31	8.5	<-avg (c)= c-bar
32	2.9155	<-c-sigma
33	17.246	<-UCL
34	0.000	<-LCL

Process Capability

The following formulas can be used to compare your actual process performance to the specification limits. These formulas assume the process is in control and data is normally distributed about a central mean or average.

$$Cp = \frac{USL - LSL}{6\hat{\sigma}}$$

$$\hat{\sigma} = \text{estimate of process standard deviation} = \frac{\bar{R}}{d_2}$$

When the Cp < 1 the process variation exceeds spec limits. When the Cp > 1 the process variation is less than spec. limits, but in either case the Cp only looks at process potential, meaning the actual mean must be centered within the spec limits to make full use of the limits with the process variation occurring.

It is stated above that Cp is process potential. Cpk is an indicator of process capability in that it looks at the variation relative to where the mean of the process is located.

$$\text{Cpk (lower)} = \frac{\bar{\bar{X}} - LSL}{3\hat{\sigma}} \qquad \text{Cpk (upper)} = \frac{USL - \bar{\bar{X}}}{3\hat{\sigma}}$$

Cpk = minimum of either {Cpk(lower),Cpk(upper)}.
Cpk values greater than 1 are favorable.
Ppk = minimum of either {Ppk(lower),Ppk(upper)} ... same as Cpk above except divisor is the calculated sample std dev (s) instead of estimated sigma as shown above.

Example:
Specification limit is 2.375 ± 0.010
The process grand average is 2.371
The r bar is 0.003; n = 4; thus, d2 is 2.09

$$\hat{\sigma} = \frac{\bar{R}}{d_2} = \frac{0.003}{2.059} = 0.001457$$

$$Cpk = \frac{\bar{\bar{X}} - LSL}{3\hat{\sigma}} = \frac{2.371 - 2.365}{3 \times 0.001457} = 1.37$$

The z distance can then be used to determine area Pz on the Z table to determine percentage of defects; it is calculated as follows. Note: The process must be in statistical control and the histogram should indicate a normal distribution.

$$Z(lo) = \frac{\bar{\bar{X}} - LSL}{1\hat{\sigma}} \quad \& \quad Z(hi) = \frac{USL - \bar{\bar{X}}}{1\hat{\sigma}}$$

$$Z(lo) = \frac{2.371 - 2.365}{0.001457} = 4.11$$

4.11 yields zero defects (per table on next page, but 6 sigma tables would indicate a Z of 4.11 to be 20 ppm defects!), but if r bar had been 0.005 then the std dev would have been 0.002428 and Z would be 2.47 which would give an area Pz of 0.68% defects (0.0068). We should calculate Z(hi) and Z(lo) in cases of large variation (range).

Z Table for Process Capability Calculations

PZ = THE PROPORTION OF PROCESS OUTPUT BEYOND A SINGLE SPECIFICA-
TION LIMIT THAT IS Z STANDARD DEVIATIONS UNITS AWAY FROM THE
PROCESS AVERAGE (FOR A PROCESS IN STATISTICAL CONTROL AND
NORMALLY DISTRIBUTED).

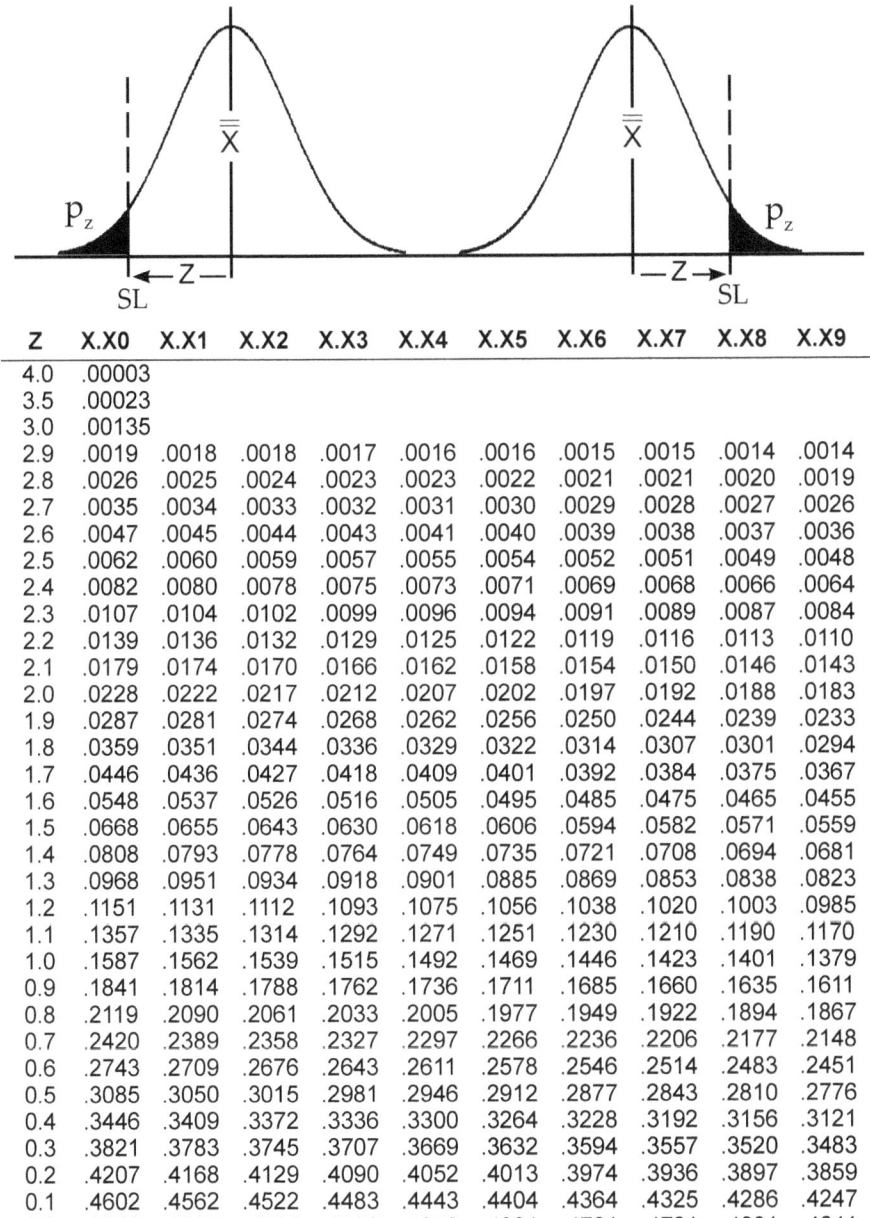

Z	X.X0	X.X1	X.X2	X.X3	X.X4	X.X5	X.X6	X.X7	X.X8	X.X9
4.0	.00003									
3.5	.00023									
3.0	.00135									
2.9	.0019	.0018	.0018	.0017	.0016	.0016	.0015	.0015	.0014	.0014
2.8	.0026	.0025	.0024	.0023	.0023	.0022	.0021	.0021	.0020	.0019
2.7	.0035	.0034	.0033	.0032	.0031	.0030	.0029	.0028	.0027	.0026
2.6	.0047	.0045	.0044	.0043	.0041	.0040	.0039	.0038	.0037	.0036
2.5	.0062	.0060	.0059	.0057	.0055	.0054	.0052	.0051	.0049	.0048
2.4	.0082	.0080	.0078	.0075	.0073	.0071	.0069	.0068	.0066	.0064
2.3	.0107	.0104	.0102	.0099	.0096	.0094	.0091	.0089	.0087	.0084
2.2	.0139	.0136	.0132	.0129	.0125	.0122	.0119	.0116	.0113	.0110
2.1	.0179	.0174	.0170	.0166	.0162	.0158	.0154	.0150	.0146	.0143
2.0	.0228	.0222	.0217	.0212	.0207	.0202	.0197	.0192	.0188	.0183
1.9	.0287	.0281	.0274	.0268	.0262	.0256	.0250	.0244	.0239	.0233
1.8	.0359	.0351	.0344	.0336	.0329	.0322	.0314	.0307	.0301	.0294
1.7	.0446	.0436	.0427	.0418	.0409	.0401	.0392	.0384	.0375	.0367
1.6	.0548	.0537	.0526	.0516	.0505	.0495	.0485	.0475	.0465	.0455
1.5	.0668	.0655	.0643	.0630	.0618	.0606	.0594	.0582	.0571	.0559
1.4	.0808	.0793	.0778	.0764	.0749	.0735	.0721	.0708	.0694	.0681
1.3	.0968	.0951	.0934	.0918	.0901	.0885	.0869	.0853	.0838	.0823
1.2	.1151	.1131	.1112	.1093	.1075	.1056	.1038	.1020	.1003	.0985
1.1	.1357	.1335	.1314	.1292	.1271	.1251	.1230	.1210	.1190	.1170
1.0	.1587	.1562	.1539	.1515	.1492	.1469	.1446	.1423	.1401	.1379
0.9	.1841	.1814	.1788	.1762	.1736	.1711	.1685	.1660	.1635	.1611
0.8	.2119	.2090	.2061	.2033	.2005	.1977	.1949	.1922	.1894	.1867
0.7	.2420	.2389	.2358	.2327	.2297	.2266	.2236	.2206	.2177	.2148
0.6	.2743	.2709	.2676	.2643	.2611	.2578	.2546	.2514	.2483	.2451
0.5	.3085	.3050	.3015	.2981	.2946	.2912	.2877	.2843	.2810	.2776
0.4	.3446	.3409	.3372	.3336	.3300	.3264	.3228	.3192	.3156	.3121
0.3	.3821	.3783	.3745	.3707	.3669	.3632	.3594	.3557	.3520	.3483
0.2	.4207	.4168	.4129	.4090	.4052	.4013	.3974	.3936	.3897	.3859
0.1	.4602	.4562	.4522	.4483	.4443	.4404	.4364	.4325	.4286	.4247
0.0	.5000	.4960	.4920	.4880	.4840	.4801	.4761	.4721	.4681	.4641

Z Table for Values at Z = 3 to 6

Z	X.X0	X.X1	X.X2	X.X3	X.X4	X.X5	X.X6	X.X7	X.X8	X.X9
3	0.001349898	0.001306238	0.001263873	0.001222769	0.001182891	0.001144207	0.001106685	0.001070294	0.001035003	0.001000782
3.1	0.000967603	0.000935437	0.000904255	0.000874032	0.000844739	0.000816352	0.000788846	0.000762195	0.000736375	0.000711364
3.2	0.000687138	0.000663675	0.000640953	0.000618951	0.000597648	0.000577025	0.000557061	0.000537737	0.000519035	0.000500937
3.3	0.000483424	0.000466480	0.000450087	0.000434230	0.000418892	0.000404058	0.000389712	0.000375841	0.000362429	0.000349463
3.4	0.000336929	0.000324814	0.000313106	0.000301791	0.000290857	0.000280293	0.000270088	0.000260229	0.000250707	0.000241510
3.5	0.000232629	0.000224053	0.000215773	0.000207780	0.000200064	0.000192616	0.000185427	0.000178491	0.000171797	0.000165339
3.6	0.000159109	0.000153099	0.000147302	0.000141711	0.000136319	0.000131120	0.000126108	0.000121275	0.000116617	0.000112127
3.7	0.000107800	0.000103630	0.000099611	0.000095740	0.000092010	0.000088417	0.000084957	0.000081624	0.000078414	0.000075324
3.8	0.000072348	0.000069483	0.000066726	0.000064072	0.000061517	0.000059059	0.000056694	0.000054418	0.000052228	0.000050122
3.9	0.000048096	0.000046148	0.000044274	0.000042473	0.000040741	0.000039076	0.000037475	0.000035936	0.000034458	0.000033037
4	0.000031671	0.000030359	0.000029099	0.000027888	0.000026726	0.000025609	0.000024536	0.000023507	0.000022518	0.000021569
4.1	0.000020658	0.000019783	0.000018944	0.000018138	0.000017365	0.000016624	0.000015912	0.000015230	0.000014575	0.000013948
4.2	0.000013346	0.000012769	0.000012215	0.000011685	0.000011176	0.000010689	0.000010221	0.000009774	0.000009345	0.000008934
4.3	0.000008540	0.000008163	0.000007801	0.000007455	0.000007124	0.000006807	0.000006503	0.000006212	0.000005934	0.000005668
4.4	0.000005413	0.000005169	0.000004935	0.000004712	0.000004498	0.000004294	0.000004098	0.000003911	0.000003732	0.000003561
4.5	0.000003398	0.000003241	0.000003092	0.000002949	0.000002813	0.000002682	0.000002558	0.000002439	0.000002325	0.000002216
4.6	0.000002112	0.000002013	0.000001919	0.000001828	0.000001742	0.000001660	0.000001581	0.000001506	0.000001434	0.000001366
4.7	0.000001301	0.000001239	0.000001179	0.000001123	0.000001069	0.000001017	0.000000968	0.000000921	0.000000876	0.000000834
4.8	0.000000793	0.000000755	0.000000718	0.000000683	0.000000649	0.000000617	0.000000587	0.000000558	0.000000530	0.000000504
4.9	0.000000479	0.000000455	0.000000433	0.000000411	0.000000391	0.000000371	0.000000352	0.000000335	0.000000318	0.000000302
5	0.000000287	0.000000272	0.000000258	0.000000245	0.000000233	0.000000221	0.000000210	0.000000199	0.000000189	0.000000179
5.1	0.000000170	0.000000161	0.000000153	0.000000145	0.000000137	0.000000130	0.000000123	0.000000117	0.000000111	0.000000105
5.2	0.000000100	0.000000094	0.000000089	0.000000085	0.000000080	0.000000076	0.000000072	0.000000068	0.000000065	0.000000061
5.3	0.000000058	0.000000055	0.000000052	0.000000049	0.000000046	0.000000044	0.000000042	0.000000039	0.000000037	0.000000035
5.4	0.000000033	0.000000032	0.000000030	0.000000028	0.000000027	0.000000025	0.000000024	0.000000023	0.000000021	0.000000020
5.5	0.000000019	0.000000018	0.000000017	0.000000016	0.000000015	0.000000014	0.000000013	0.000000013	0.000000012	0.000000011
5.6	0.000000011	0.000000010	0.000000010	0.000000009	0.000000009	0.000000008	0.000000008	0.000000007	0.000000007	0.000000006
5.7	0.000000006	0.000000006	0.000000005	0.000000005	0.000000005	0.000000004	0.000000004	0.000000004	0.000000004	0.000000004
5.8	0.000000003	0.000000003	0.000000003	0.000000003	0.000000003	0.000000002	0.000000002	0.000000002	0.000000002	0.000000002
5.9	0.000000002	0.000000002	0.000000002	0.000000002	0.000000001	0.000000001	0.000000001	0.000000001	0.000000001	0.000000001
6	0.00000000099									

Nominalization

Page 62 mentions that Cpk & Ppk look at where the process mean is located when computing the capability number. There are three main variables comprising the Cpk number that describe our dimensional compliance for molded products. The three main variables in this equation are as follows:

1. standard deviation – in this case sigma hat which is an estimated standard deviation based on R bar ÷ d_2 from control chart data. When a true sample sigma (pooled standard deviation) is calculated and used the Cpk is referred to as Ppk.
2. mean or grand average of parts measured.
3. specification limits: LSL & USL.

Of the three aforementioned components to the Cpk calculation, we as custom molders and engineers have control over two of the three: standard deviation and the process mean. The customer controls the specification limits.

The standard deviation indicates how much variation there is in the process. This includes all sources of variation including measurement, core/cavity size variation, process variation caused by resin and machine variation, etc. The formulas below quantify the mean relative to specification limits with a normalized zero-to-one scale. In the graphic below, the nominalization is 0.85 (85%).

Nominalization Calculation Quantifies Process Centering

$$1 - \left\{ \frac{(target - mean)}{(target - LSL)} \right\} = c \ (centering)$$

If mean is closer to USL...

then use the following equation

$$1 - \left\{ \frac{(mean - target)}{(USL - target)} \right\}$$

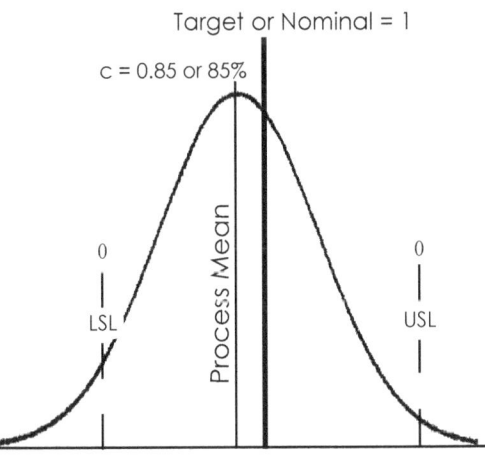

Target or Nominal = 1

c = 0.85 or 85%

Process Mean

0 — LSL

0 — USL

A perfectly centered process has a nominalization of 1 (100%); the closer the mean gets to the specification limits, the nominalization becomes zero. In this example, the nominalization is 0.85 (85%) ... (see formulas above).

This same formula converted to simple spreadsheet format would look like this:
=IF(mean>target,1-(mean-target)/(USL-target),1-(target-mean)/(target-LSL)).

Cpk vs. Z Scores

The Z score computation is <u>nearly identical</u> to Cpk/Ppk except we divide by one std dev instead of three; henceforth, a Cpk of two is the same as a Z score of six. This seemingly simple relationship does get "clouded" when one enters the world of six sigma quality control. In the 6σ world, Z scores are typically used instead of Cpk or Ppk. There are tables to convert Z score to ppm defective. In Excel and other spreadsheet software there are functions to convert Z scores into ppm defective. The formulas are so similar, we could multiply a Cpk/Ppk score by three and then use the standard Excel function to compute ppm defective.

$$\text{Cpk (lower)} = \frac{\overline{\overline{X}} - LSL}{3\hat{\sigma}} \qquad \text{Cpk (upper)} = \frac{USL - \overline{\overline{X}}}{3\hat{\sigma}}$$

$$Z_{lsl} = \frac{\overline{\overline{X}} - LSL}{1\,\sigma_{n-1}} \quad \text{and} \quad Z_{usl} = \frac{USL - \overline{\overline{X}}}{1\,\sigma_{n-1}}$$

In order to properly identify the <u>total</u> ppm defective, we would need to add the ppm defective from Z_{LSL} and from Z_{USL}. We could do the same for Cpk upper and lower, but unfortunately the Cpk scores are <u>typically</u> computed and reported as a single number; thus, <u>given as a single Cpk score</u>. One must then assume that the process was centered whereby the defects in the tail(s) of the bell curve, and beyond specification limits, occur evenly on both sides; thus, we double the ppm computed. The computation is frequently done in spreadsheet software such as Excel and is shown below.

$ppm = (1\text{-normsdist}\,(C_{pk}{*}3))\,x\,2\,x\,1000000$

$ppm = (1\text{-normsdist}\,(Z_{LSL}))\,x\,1000000 + (1\text{-normsdist}\,(Z_{USL}))\,x\,1000000$

The difference between the two is the factor of 3 for Cpk because Cpk calculation has a divisor of 3. We multiply by 2 to assume even tails or good process centering. Identical ppm defective could be achieved if we keep track of the Cpk upper and lower just as we do for Z scores.

Cpk vs. Z Scores ... 1.5 Sigma Shift

The aforementioned "clouded" reference relates to a 1.5σ shift allowance that 6σ black belts would include (6σ is really 4.5σ in order to allow for long term data shift and drift ... predicted via calculation to be 1.5σ ... thus, it includes a 1.5 shift ... i.e. 6 includes a 1.5 addition from 4.5). In order to convert from long term data back to short term data: subtract 1.5.

If data is long term then add or subtract 1.5 as follows (assumes an official Z is based on long term data): ppm defective = 1000000 * (1-NORMSDIST(Z-1.5)) and Z = ((NORMSINV (ppm/1000000)*-1)+1.5). NOTE: Spreadsheets thruout this book do not include any 1.5 sigma shifts as data is short term in nature. When converting long term Z to short term Z, subtract 1.5 .... when converting short term Z to long term Z, add 1.5.

Any time we calculate a Z-score based on PPM from a sampling of data, this is short term data without long term shifts and drifts. We can convert this using a Z table whereby 3.4 ppm is 0.0000034 decimal equivalent. Look up 0.0000034 in Z table and you find a Z value of 4.5. We can also convert the 3.4 ppm back to Z value w/ EXCEL® function = (NORMSINV(3.4/1000000)*-1) and get Z of 4.5 ... we can convert the Z of 4.5 back to ppm w/EXCEL® function =1000000*(1-NORMSDIST(4.5)) and get 3.4 ppm; thus, the calculations are related to each other and do not need 1.5 added or subtracted to be relative to each other.

In graphic on previous page and reshown below, the nominalization is 0.85; if the variation was very large, then there could be <u>discrepant product beyond both speci-</u><u>fication limits</u>. The typical Cpk calculation would focus only on the LSL since more defects would occur beyond that limit.

If we compute ppm defective for both tails and then add the two ppm numbers we can then convert back to a composite overall Z score (Total Z=NORMSINV(total ppm/1000000)*-1) ... Note: This Z score does not include a 1.5 sigma shift; thus you should add 1.5 to yield an official six sigma score ... assumes the data is long term or can be sustained on a long term basis.

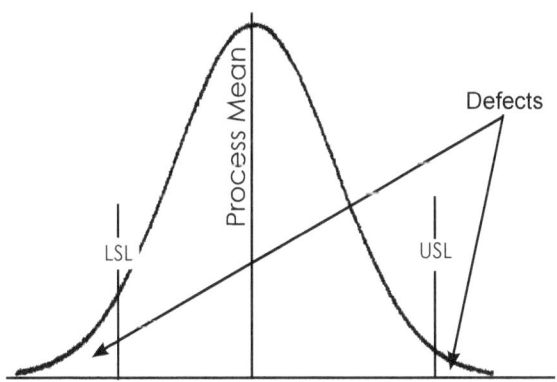

NOTE: Food for thought ... For internal use, when there is no need for metrics to impress, consider never adding or subtracting any 1.5 values to adjust sigma for long term shift or drift. If the data set has a given std dev, spec limit and mean, the calculated ppm defective and Z score will be accurate for that data set!

Z Score Spreadsheet Math

Row three in spreadsheet below could be listing data for the bell curves found on preceding pages with a nominalization of 0.85 and some defects (889 ppm) beyond the lower spec limit and a lesser number of defects beyond the upper spec limit (12 ppm) for a total of 901 ppm defective.

	A	B	C	D	E	F	G	H	I	J	K
1	Valve										
2		Units	Target	LSL	USL	Mean	Sigma	C	Z_{LSL}	Z_{USL}	PPM
3	1	inch	4.250	4.245	4.255	4.2493	0.001360	0.85	3.1250	4.23	901
4	1	inch	0.875	0.872	0.878	0.8751	0.000925	0.97	3.35	3.14	1261
5											
6											
7							TOTAL D				2162

Formulas in cells are as follows:
cell H4 =IF(NOT(ISBLANK(C4)), IF((E4-F4)<(F4-D4), 1-(F4-C4)/(E4-C4), 1-(C4-F4)/(C4-D4)),"")
cell I4 =IF(ISBLANK(D4),"",IF(G4<>0,(F4-D4)/G4,""))
cell J4 =IF(ISBLANK(E4),"",IF(G4<>0,(E4-F4)/G4,""))
cell K4 =IF(NOT(ISBLANK(C4)),1000000*(IF(ISBLANK(D4), 0, IF(G4<>0,1-NORMS DIST(I4),0))+IF(ISBLANK(E4), 0, IF(G4<>0,1-NORMSDIST(J4),0))),"")

In example spreadsheet above, the 0.875 dimension is well centered with a c = 0.97 (nominalization = 97%), but has more ppm defective. This is due to the ratio of sigma vs. tolerance. The 0.875 dimension has a tighter tolerance. The table below, will show the ppm defective for each the Z lower and upper tails. The formulas for PPM-L & PPM-U would be similar to cell K4 above, except K4 above adds both together. If we break out the misc IF statements to maintain proper display for blank cells, the PPM-U would be:

Z_{LSL}	Z_{USL}	PPM-L	PPM-U
3.12	4.23	**889.0**	**11.8**
3.35	3.14	**402.1**	**858.9**

=1000000 x (1-NORMSDIST(J3))
=11.8 (or 11.7 for 4.23 ... w/o hidden digits)

NOTE: If you try to re-create in your spreadsheet, your answer may be off slightly because of display rounding.

We can calculate a Z score for the total of 2162 pcs (assumes 901 are not the same pieces as the 1261):

= Z score for cell K7 (above)
= NORMSINV (K7/1000000) x -1 ... same as ... = NORMSINV (2162/1000000) x -1
= 2.85 (w/o 1.5 sigma shift)

Note also the 2162 is ppm defective which can also be thought of as 2162/1000000 or 0.002162 defects per unit or a 0.216 % chances of having a defect.

Take care when trying to equate Cpk values to Z scores. It may work best to compute a Cpk lower and Cpk upper, then compute PPM total by taking 1000000 x (1- Normsdist of Cpk lower x 3) + 1000000 x (1- Normsdist of Cpk upper x 3) ... this is basically converting Cpk upper and lower scores to Z scores. A single Cpk value ignores possible defects from other tail.

Process Mapping

As one begins an attempt to manage variation, it is necessary to identify the following:

1. What are the outputs (e.g. molded parts, cycle time, measured dimensions, efficiency, production yields, scrap, performance such as Cpk, Ppk, Z score, etc.)?
2. What are the inputs (e.g. resin, mold, machine, time, pressure, fill speed, mold temperature, melt temperature, gage, methods, process sheet, etc.)?

Our inputs "x" combine to yield an output y; thus, $y = f(x_1, x_2, x_3 ...)$.

It is best if the inputs and outputs are quantitative (measurable). This permits exact values, but also permits variation to be quantified, DOEs to be performed, regression analysis, etc ... subjects discussed later in this book.

Many of the inputs relate to molding machine controls or knobs; thus, some of these are controllable; whereas, some are less controllable. The variables that have no or minimal control associated can be classified as noise (e.g. people, materials). As we develop our process map, we can label the inputs as controllable or noise. We can also label some as critical once we identify which inputs are critical thru profound knowledge (DOEs work well in absence of profound knowledge).

As can be seen in graphic below, the interaction of inputs affects our output variation. The relationship is not always a straight line relationship; sometimes there are operating ranges whereby the input has a lesser effect on output variation. This concept is not uncommon to molders who understand the benefits of performing injection forward optimization studies (gate seal time) or viscosity curves to identify reduced viscosity variation at faster fill times. Some process factors may behave with more efficacy and less variability depending on other interacting process factors. The same pack pressure may have different results at different pack times!

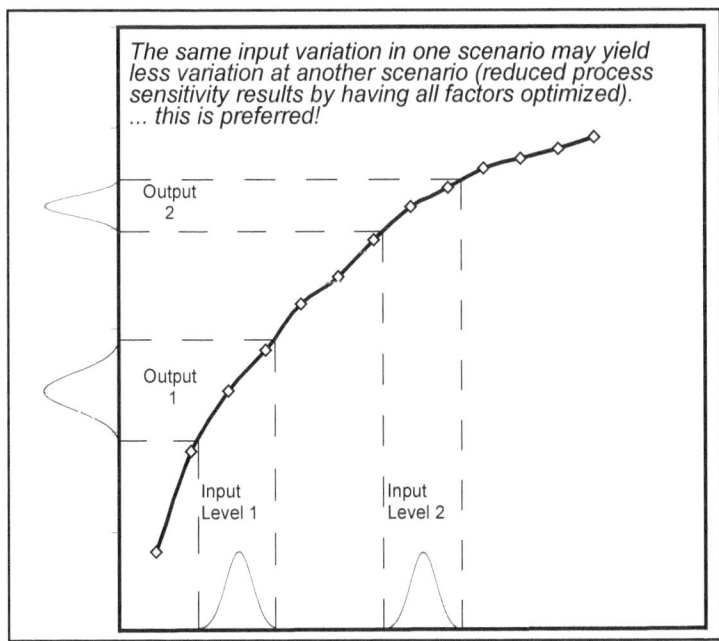

The same input variation in one scenario may yield less variation at another scenario (reduced process sensitivity results by having all factors optimized). ... this is preferred!

Process Mapping (Dryer Example)

Details of process step for drying – from process map below for injection molding. Each process step at bottom of page would have a similar detailed listing of inputs & outputs. Some inputs can be identified as critical (if proven by DOE or other means).

Walk the process to best ID all inputs/outputs (sometimes there are undocumented activities that may be important ... aka "hidden factories"). Once detailed as shown, we can then perform an FMEA or identify factors for DOEs or other improvement techniques. Attempt to make all or most I/Os quantifiable so can be measured for compliance (and used in DOEs).

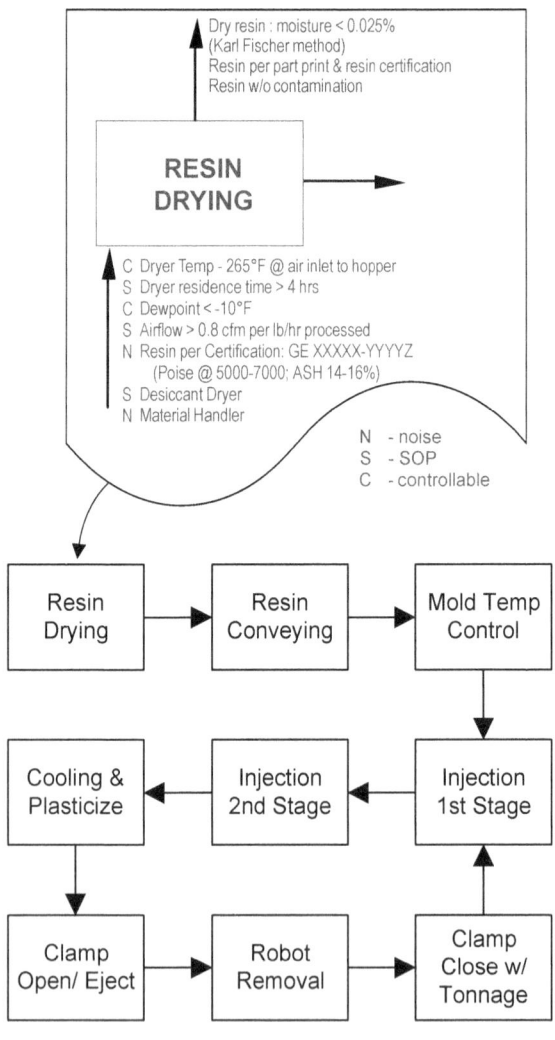

Failure Modes and Effects Analysis

Also known as FMEA, is a systematic analysis of the potential failure modes – failure to achieve the desired (required) outputs. It includes the identification of:

1. Failure modes – what is actually seen relative to the expected output from the PMAP.
2. Potential causes – typically relates to inputs on the PMAP.
3. Effects – what the customer (internal or external) will realize or see.
4. Risk analysis: ranking based on severity, occurrence and detection. A rating table is required to rate severity, likelihood of occurrence and ability to detect; the table might include three or five choices which includes a number and description.
5. List normal control methods.
6. ID items having highest risk priority number (RPN which is product of the severity, occurrence and detection rating done in #4 above) and develop an action plan to improve controls which can be prevention, reduced occurrences, better detection, etc.

This chart is frequently done in a spreadsheet format. This chart is easier to create if a PMAP has been made which identifies all of the process steps with inputs and outputs. We will want to list all the process steps and concentrate on those inputs that are ranked as critical. In the sample entry listed below, the risk priority number (RPN) has been reduced by improving detection (3 to 1). Compare the RPN to post RPN (Prpn) ... columns #8 below are all post values after severity, occurrence and detection have been improved and re-rated.

FMEA														
1	2	3	6	4	6	5	6	6	7	8	8	8	8
Part/Process	Failure Mode	Failure Effects	SEV	Causes	OCC	Current Controls	DET	RPN	Actions	PS	PO	PD	Prpn
Resin Drying	Resin not dry.	Adverse affect on viscosity resulting in over or under packed parts and possible discrepant product - flash or dimensional problems	4	Dryer residence time too short…less than 4 hrs	2	Incoming inspection compares resin cert to specification limits. Certs req'd before resin entered into inventory.	3	24	Use of RJG device which can sense the viscosity change and alarm accordingly… requires purchase & installation of eDart	4	2	1	8

Severity Rating Table
1. No known effect on molded product.
2. Minor shift in dimensions; meets control limits and spec limits.
3. Out of control limits, but within spec limits.
4. Out of spec limits, but still functional.
5. Out of spec; physical properties effected; parts not functional.

Occurrence Rating Table
1. Yearly
2. Quarterly
3. Monthly
4. Weekly
5. Daily

Detection Rating Table
1. Detected at the molding press, containment w/o scrap produced.
2. Detection at molding press w/ some scrap produced (< 1 hr).
3. Detection in molding, but after quality audit (1 - 4 hrs scrap).
4. Detection before ship (> 4 hrs scrap).
5. Defects not detected; shipped to customer.

Correlation

In order to perform real time SPC, it is useful to understand correlation and linear regression. Correlation analysis will help to identify opportunities for real time SPC – what dimensional feature can be predicted by a certain process output from the machine that can be monitored real time (ideally, real time means before the mold opens; thus, allowing a signal to robot or conveyor to reject and scrap the appropriate parts). Linear regression can be used to verify statistical significance of the real time process parameter, and to create a math equation for use in calculating alarm limits and/or to attempt to fully process correct dimensional deficiencies with process corrections.

The formula below is reprinted from the Math Skills book for Injection Molding, 2nd Ed. The objective of such a calculation is to test for strong direct or indirect proportional relationships. The formula below measures such correlation strength on a scale of minus one to one (-1 to 1) whereby 1 is a perfect direct correlation and -1 is a perfect indirect correlation, and a value near zero(0) has little or no correlation (the closer we get to 1 and -1; the stronger the correlation).

$$r = \frac{\sum X_i Y_i - n\overline{X}\ \overline{Y}}{(n-1)s_x s_y}$$

Sx & Sy are sample standard deviations of X and Y samples. The function CORREL in Excel will derive the same answer as formula above. In data set on next page, we see the opportunity to perform real time SPC by monitoring C1_Peak because of it's 0.91 correlation to L-avg. We decide we will set limits whereby we do not want to have lengths lower than 4.222 or higher than 4.228 (0.002 inside of spec. limits). We also decide we will make a steel adjustment of 0.005 inches so we improve our nominalization and avoid excessive pressures needed to process correct this dimension. In order to determine the predictive equation for what pressures yield L-avg of 4.222 & 4.228, we will need to understand linear regression analysis discussed later.

Correlation ... Spreadsheet Example

DOE Data Example #2
(correlation of process data to part quality data)

	A	B	C	D	E	F	G	H	I	J	K	L
1					Pack	Mold	Fill	control dimensions				
2	SO	RO	CP	BLK	psi	Temp	Time	L - avg	H - avg	W - avg	C1_C.I.	C1_Peak
3	1	8	1	1	5500	100	0.60	4.2148	0.8722	1.5015	12416	4612
4	2	9	1	1	7000	100	0.60	4.2235	0.8732	1.5018	15612	5611
5	3	4	1	1	5500	120	0.60	4.2155	0.8725	1.5016	12878	4684
6	4	5	1	1	7000	120	0.60	4.2276	0.8750	1.5015	16245	5916
7	5	2	1	1	5500	100	1.00	4.2142	0.8711	1.5016	12156	4580
8	6	3	1	1	7000	100	1.00	4.2189	0.8729	1.5016	15410	5404
9	7	6	1	1	5500	120	1.00	4.2198	0.8725	1.5016	13210	4788
10	8	7	1	1	7000	120	1.00	4.2236	0.8741	1.5018	15890	5712
11	9	1	0	1	6250	110	0.80	4.2219	0.8735	1.5017	14112	5068
12	correl							0.89	0.86	0.41	**1.00**	
13	correl							0.91	0.87	0.37		**1.00**

`=CORREL($L3:$L110, H3:H11)`

In spreadsheet above, the rows for "correl" correlate each column of data to the column with the perfect 1.00 in BOLD (1.00 because that column is compared to itself; thus, perfect 1.00...useful to ID basis for comparison and worksheet construction to copy cells with formulas across).

Linear Regression Analysis

If we perform regression analysis on this data set of L-avg vs C1_Peak (cav psi), an equation is developed that describes the best fit line shown in graph below ... equation is as follows:

$$\text{L-avg} = 4.17850 + 0.000000805 \times \text{C1\_Peak}$$

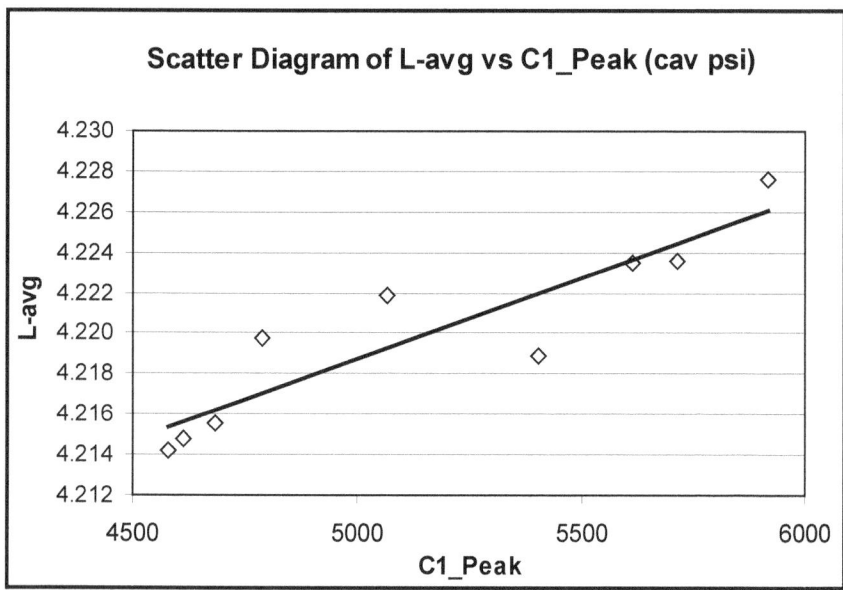

The graph above is a simple XY (scatter) plot from Excel®; whereas, the data output on next page is from Minitab®. Either software can generate this plot (or closely similar) and the data output on next page. Minitab does offer many more tools when it comes to statistical data analysis and graphical display thereof ... both examples will be shown to offer alternatives for some more basic data analysis.

Linear Regression Analysis: Minitab®

For best accuracy substitute the coefficient values seen below "Coef" column into your equation

```
Regression Analysis: L - avg versus C1_Peak
The regression equation is
L - avg = 4.18 + 0.000008 C1_Peak

Predictor        Coef      SE Coef        T      P
Constant      4.17850      0.00725   576.18  0.000
C1_Peak     0.00000805   0.00000140     5.74  0.001

S = 0.00205389   R-Sq = 82.5%   R-Sq(adj) = 80.0%

Analysis of Variance
Source          DF         SS          MS       F      P
Regression       1  0.00013923  0.00013923   33.00  0.001
Residual Error   7  0.00002953  0.00000422
Total            8  0.00016876

Obs  C1_Peak   L - avg       Fit   SE Fit   Residual   St Resid
  1     4612   4.21480   4.21562  0.00102   -0.00082      -0.46
  2     5611   4.22350   4.22367  0.00094   -0.00017      -0.09
  3     4684   4.21550   4.21620  0.00095   -0.00070      -0.39
  4     5916   4.22760   4.22612  0.00127    0.00148       0.92
  5     4580   4.21420   4.21537  0.00105   -0.00117      -0.66
  6     5404   4.21890   4.22200  0.00077   -0.00310      -1.63
  7     4788   4.21980   4.21704  0.00085    0.00276       1.48
  8     5712   4.22360   4.22448  0.00104   -0.00088      -0.50
```

from Minitab® Statistical Software, release 16.1

NOTE: Take care when extracting data from Minitab® output so that all significant digits are used in the predictive equation ... the equation is rounding data taken from the regression coefficients; use actual coefficients for best accuracy. Different regression selection options in Minitab® will yield different displays of these digits ... some options display up to 8 decimal places and some display 11 places (w/ scientific notation). This extra data resolution may not be needed for your data set, but never hurts!

The R-Sq value of 82.5% agrees with the r value of 0.91 found on page 77 – Excel® analysis ($0.91^2 = 0.825$). The R-Sq indicates how much variation is explained. The analysis of variance proves the data is statistically significant. The F value of 33.00 is derived by taking the MS (mean square) for the regression and dividing by MS for residual error ($0.00013923 \div 0.00000422$). The P value is the probability that the same result could occur from chance. Higher F values (or T) and lower P values ($P<0.05$) are good. When there are multiple regressor terms, the R-Sq adjusted value is a better indicator of variation explained.

Linear Regression Analysis Using Excel®

The predictive equation can also be derived in Excel; turn needed feature on by selecting: TOOLS, ADD-INS, ANALYSIS TOOL PACK, OK...then select: TOOLS, DATA ANALYSIS, REGRESSION.

It matters how you fill in the menu at right: think of the Input X range as Y= function of X or L-avg is a function of C1_Peak. Note: The cell references are for cells shown in DOE Data Ex. #2, (earlier pages).

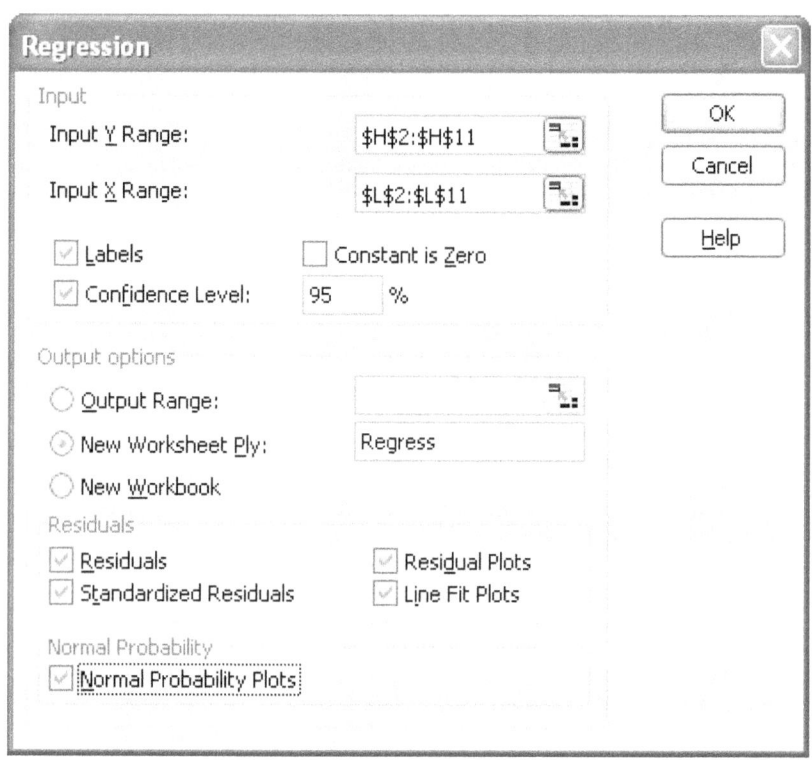

You will get an output that has much the same information as MINITAB®, but you have to understand that the format for linear regression math equations: $y = c + b_1x$ ... (technically there is also an error term to be added). It should also be pointed out that our processes have many inputs; thus, for a given output y, the equation may be more like: $y = c + b_1x_1 + b_2x_2 \ldots + b_nx_n + E$. It is important and valuable to understand the process sufficient to identify the main inputs effecting output Y ... and to minimize the error or noise values effecting output y.

Linear Regression ... Excel® Spreadsheet Output

SUMMARY OUTPUT

Regression Statistics	
Multiple R	0.908304586
R Square	0.82501722
Adjusted R Square	0.80001968
Standard Error	0.002053892
Observations	9

ANOVA

	df	SS	MS	F	Signif F
Regression	1	0.000139226	0.000139226	33.004	0.000702082
Residual	7	2.95293E-05	4.21847E-06		
Total	8	0.000168756			

	Coeff	Std Err	t Stat	P-value	Lower 95%
Intercept	4.178501443	0.007252061	576.1812236	1E-17	4.161353056
C1_Peak	8.04932E-06	1.40112E-06	5.744905262	0.0007	4.73619E-06

RESIDUAL OUTPUT

Observation	Pred L-avg	Residuals	Std Residuals
1	4.21562	-0.00082	-0.429351
2	4.22367	-0.00017	-0.086482
3	4.21620	-0.00070	-0.366658
4	4.22612	0.00148	0.769714
5	4.21537	-0.00117	-0.607581
6	4.22200	-0.00310	-1.613512
7	4.21704	0.00276	1.435757
8	4.22448	-0.00088	-0.457587
9	4.21930	0.00260	1.355700

from Microsoft EXCEL®

$y = c + b_1 x$

c is the constant or intercept (intercept coeff as shown above).
b_1 is the C1_Peak coeff.
x is the C1_Peak value used ... such as 5200 psi (peak cav psi).

Multiple regression can have multiple predictors ($b_1 x_1$, $b_2 x_2$, ...).

Excel output above is partial to save space. As can be seen, the coefficients are the same as derived in Minitab®: 4.1785 & 0.00000805.

The r^2 value indicates how much variation is explained ... i.e. the C1_Peak variation accounts for 82.5% of the L-avg variation. This is strong, because peak psi is a main determinant in molded part sizing for a given cavity size. The residuals are the difference between actual data points and the line represented by math equation. Outliers can effect the strength of model; thus, large std residuals need to be reviewed w/ decision for inclusion to model. This emphasizes need for good experimental practices AND need for performing MSE.

Calculating Alarm Limits for SPC

From a previous page, we used linear regression to yield this formula:

$$L\text{-avg} = 4.17850 + 0.000000805 \times C1\_Peak$$

however, after reviewing the data, we determined that we need to adjust steel by 0.005 inches to improve our centering. If we do not adjust steel, we must process correct the dimension which has the following potential problems:
- other dimensions could go up as well (maybe out of spec)
- excessive pressure might be required
- we would be extrapolating data beyond our inference space which may or may not be valid in that we do not know if the linear relationship extends that far beyond the basis for the predictive equation

Our alarm limit calculation is simply a rearrangement of the equation at top of page (basic algebra), AND we add 0.005 to our constant (intercept) of 4.1785 (if we did not need a steel adjustment we could ignore the 0.005 adder).

$$(\text{Lo alarm}) \ C1\_peak = \frac{\left(4.222 - \left(4.1785 + 0.005\right)\right)}{0.00000805}$$

$$= \frac{\left(4.222 - 4.1835\right)}{0.00000805}$$

$$= \frac{0.0385}{0.00000805}$$

$$= 4783 \text{ psi}$$

The Hi alarm limit chosen to signal > 4.228 would be computed in similar fashion (is 5528 psi). This can be programmed into Excel® (shown below). Here we see we want to maintain a cavity psi of 5155 psi and set alarm limits at 4783 and 5528 (4.222 - 4.228). This approach equates alarm values to specific dimensions rather than just using the ± 4.5σ described earlier.

	nominal	low alarm limit	hi alarm limit
target	4.225	4.222	4.228
coeff	4.1785	4.1785	4.1785
coeff	0.00000805	0.00000805	0.00000805
predicted before steel adj	**4.1785**	**4.1785**	**4.1785**
steel adj >>	0.005	0.005	0.005
calc C1_peak	**5155**	**4783**	**5528**

NOTE: This does assume all cavities behave in a similar fashion. An Excel plot of all cavities can test for correlation between cavities (similar behavior). If cavities are higher and lower due to imbalanced fills or steel sizing differences, then the alarm limits can be adjusted accordingly. For maximum assurance of good quality, pressure transducers could be placed in each cavity.

In the event transducers do not exist, then correlation and regression can still be done on machine process monitored parameters such as boost cutoff PSI, pack/hold psi, fill time, mold temperature, cushion, etc as needed to see what might correlate well to the molded part quality parameter of interest.

Multi Variable Regression

As stated on a previous page there can be multiple regression variables (e.g. b_1x_1, b_2x_2, ...). There will still be a constant or intercept plus coefficients for the various process factors.

With multiple regressors, the R^2 (adj) becomes important to reference in terms of identifying variation explained. If we included many regression variables, the R^2 value would approach 99% (every regressor added increases the R^2 value even if it has no effect). The R^2 (adj) looks at the degrees of freedom[1] to account for multiple factors present.

One of the many beneficial uses for multi variable regression analysis in injection molding is to combine results of DOEs with actual production data. For example: we perform a DOE and we learn that pack psi, melt temperature and resin viscosity are all important. We develop a predictive equation for these three variables, but we decide the DOE only looked at two lots of resin and that we want to have many more lots reflected in the predictive equation. We can take the responses and factors from the DOE and combine them with subsequent production data to yield a more robust regression equation (predictive equation).

Source	Date	Pack	Melt	Viscosity	Length
DOE	7/16/2002	6000	535	5318	3.559
DOE	7/16/2002	8000	535	5318	3.561
DOE	7/16/2002	6000	555	5318	3.562
DOE	7/16/2002	8000	555	5318	3.564
DOE	7/16/2002	6000	535	6120	3.557
DOE	7/16/2002	8000	535	6120	3.558
DOE	7/16/2002	6000	555	6120	3.562
DOE	7/16/2002	8000	555	6120	3.562
PROD	8/23/2002	7000	545	5511	3.561
PROD	9/13/2002	7000	545	6644	3.559
PROD	10/22/2002	7000	545	5112	3.563
PROD	11/4/2002	7000	545	5846	3.561
PROD	11/20/2002	7000	545	6222	3.560
PROD	12/12/2002	7000	545	6108	3.561

On the following page, the regression equation is derived. It should be noted that this equation is for the specific inference space tested: bounded by viscosities of 5112 - 6644; pack pressures of 6000 - 8000 psi and melt temperatures of 535 - 555 °F.

In the example on next page, the what if scenario calculates what to do when the vlscosity drops to 4750. Since this has never been tested (in the data set driving the DOEs and multi regression analysis), the derived answer is an extrapolation beyond what has been seen ... meaning the derived answer may or may not work, but can be tried. The results can be added to the data set above to yield a better regression equation for the future.

The regression analysis on next page indicates the melt temperature is more signifi- cant; thus, may be a better control or adjustment for the change in viscosity (pack is a faster adjustment, and often more significant, but the data says melt is more signifi- cant based on this data).

---

[1] Total factors (n) minus one: n -1 = number of comparisons available to estimate a statistic desired (if we had 4 straws, it would take 3 comparisons to find shortest).

Multi Variable Regression Analysis

Regression Analysis (MINITAB®): Length versus Pack, Melt, Viscosity

```
The regression equation is
Length = 3.47 +0.000001 Pack +0.000187 Melt -0.000002 Viscosity

Predictor          Coef      SE Coef          T       P
Constant        3.46632      0.01382     250.87   0.000
Pack         0.00000062   0.00000025       2.52   0.030
Melt         0.00018750   0.00002476       7.57   0.000
Viscosit    -0.00000210   0.00000041      -5.09   0.000


S = 0.0007004    R-Sq = 90.0%     R-Sq(adj) = 86.9%


Analysis of Variance

Source             DF           SS           MS          F       P
Regression          3  0.000043951  0.000014650      29.86   0.000
Residual Error     10  0.000004906  0.000000491
Total              13  0.000048857

Source        DF       Seq SS
Pack           1  0.000003125
Melt           1  0.000028125
Viscosit       1  0.000012701
```

Note 1: Use the actual listed coefficients from table below the regression equation above in subsequent calculations for optimum accuracy (MINITAB® does some rounding as shown).

We can rearrange the regression equation (w/ more accurate coefficients from table) to solve for either melt or pack as a compensator for a resin viscosity change down to 4750 poise as follows (Target = 3.562):

If pack adjusted (melt held constant at 545° F):

$$= \frac{(3.562 - 3.46632 - (0.0001875 \times 545) - (-0.0000021 \times 4750))}{0.00000062}$$

$= 5593$ psi

If melt adjusted (pack held constant at 7000 psi):

$$= \frac{(3.562 - 3.46632 - (0.00000062 \times 7000) - (-0.0000021 \times 4750))}{0.0001875}$$

$= 540°$ F

Multi Variable Regression Analysis (continued)

This could be programmed into a spreadsheet such as EXCEL and calculated as shown below:

	A	B
1	Target >>	3.562
2	melt >>	545
3	viscosity >>	4750
4	pack >>	5593
5		
6	Target >>	3.562
7	melt >>	540
8	viscosity >>	4750
9	pack >>	7000

The formulas for dependent variables pack and melt (shaded cells above) ... are shown below:

cell B4 = (B1-3.46632-(0.0001875*B2) - (-0.0000021*B3))/0.00000062
cell B7 = (B6-3.46632-(0.00000062*B9) - (-0.0000021*B8))/0.0001875

Note 2: The R^2 (adj) shown on previous page is high because we have already done screening DOEs (not shown) to determine that these variables are significant, and some variables which could be forced into significance are set as constants (i.e. pack time, cooling time).

Residual Analysis

Residuals are the difference between the observed values and predicted or fitted values; review in search of possible bad data (data that is suspect or at least unexpected).

The following multi-factor regression analysis is same as previous pages except the measured length for DOE SO #2 was originally measured at 3.564. The following regression analysis indicated that obs #2 does not fit versus what is expected: compare length below for Obs 2 at 3.564 vs Fit of 3.56168. The right hand column marks the data with an R value indicating a large residual value. The previous pages have a corrected value of 3.561 for the analysis; note how it effects the equation and significance of factors AND the R^2 (adj) value indicating how much variation has been explained.

Regression Analysis (MINITAB®): Length versus Pack, Melt, Viscosity

```
The regression equation is
Length = 3.49 + 0.000001 Pack + 0.000150 Melt - 0.000003 Viscosity


Predictor         Coef       SE Coef        T       P
Constant       3.48726       0.02287   152.50   0.000
Pack         0.00000100    0.00000041     2.44   0.035
Melt         0.00015000    0.00004098     3.66   0.004
Viscosity   -0.00000260    0.00000068    -3.81   0.003


S = 0.00115921   R-Sq = 77.2%   R-Sq(adj) = 70.4%


Analysis of Variance
Source            DF          SS           MS       F       P
Regression         3   0.000045491   0.000015164   11.28   0.002
Residual Error    10   0.000013438   0.000001344
Total             13   0.000058929


Source       DF       Seq SS
Pack          1   0.000008000
Melt          1   0.000018000
Viscosity     1   0.000019491


Obs  Pack   Length     Fit    SE Fit   Residual   St Resid
  1  6000   3.55900  3.55968  0.00073  -0.00068     -0.76
  2  8000   3.56400  3.56168  0.00073   0.00232      2.59R
  3  6000   3.56200  3.56268  0.00073  -0.00068     -0.76
  4  8000   3.56400  3.56468  0.00073  -0.00068     -0.76
  5  6000   3.55700  3.55760  0.00069  -0.00060     -0.64
  6  8000   3.55800  3.55960  0.00069  -0.00160     -1.72
  7  6000   3.56200  3.56060  0.00069   0.00140      1.51
  8  8000   3.56200  3.56260  0.00069  -0.00060     -0.64
  9  7000   3.56100  3.56168  0.00037  -0.00068     -0.62
 10  7000   3.55900  3.55873  0.00065   0.00027      0.28
 11  7000   3.56300  3.56272  0.00056   0.00028      0.28
 12  7000   3.56100  3.56081  0.00031   0.00019      0.17
 13  7000   3.56000  3.55983  0.00042   0.00017      0.16
 14  7000   3.56100  3.56013  0.00037   0.00087      0.80
R denotes an observation with a large standardized residual
```

Note: The length with high residual is not any higher than other data in set, but higher than expected given those process conditions (high pressure, but lower melt temp and lower viscosity).

Residual Plots: Minitab®

In cases with high residual values; investigate to determine root cause. The following may be the cause for such results:

1. data entry errors
2. measurement error
3. poor experimental DOE techniques (e.g. lack of process equilibrium achieved prior to collecting parts)
4. unknown variation caused by equip problems, part damage, etc.
5. other unknown variation: review PMAP, FMEA, etc

Sometimes the residuals plots (such as shown below) can help to identify outliers as well. In plot below there is no discernible pattern, but frequently the residuals vs run order is handy to see if unplanned/uncontrolled process changes have crept into the data. The "suspect" data is circled in plot below.

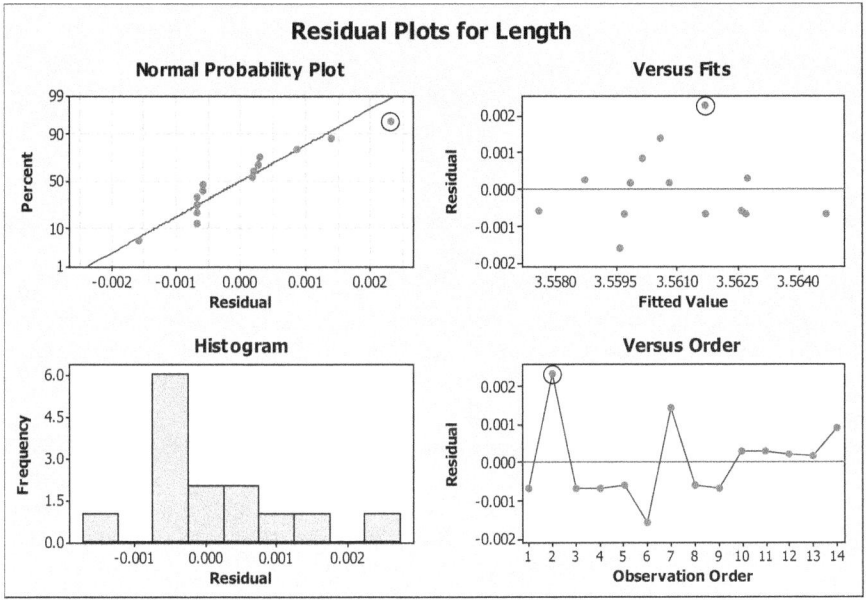

Minitab® Residual Plot: Four in One

DIII Processing

Decoupled molding accomplishes a fast fill, then performs VPT[1] (velocity control to pressure control transfer) resulting in a pack stage. This often results in a momentary stalling of the injection forward process and can sometimes result in ram bounce. This is reverse or backwards travel of the screw, but does not mean there is no pressure ahead of screw, the stored pressure ahead of screw just overpowers the hydraulics, but this still takes milliseconds of time to stabilize whereby there may be less control in the pack phase of the molding process.

An alternative is not performing VPT[1] until cavity is completely full and packed – aka Decoupled III[2]. In DIII processing we fill and pack on first stage with an abundance of pressure because the machine is in velocity control using only pressure required for the set fill speed. The only problem is that we have an abundance of pressure and run the risk of overpacking unless we monitor cavity pressure with cavity pressure transducers, and use same for the trigger signal to perform VPT.

With DIII molding, cavity pressure transducers are the preferred method for VPT. The typical setup includes a post gate transducer (near gate, but after gate) to perform this VPT and a separate transducer near end of fill to monitor or indicate part quality (after correlation analysis).

The setup for DIII molding includes filling fast as done in normal decoupled molding (DII) up to approximately 88-94% full, then slow fill speed down to approximately 10-15% of previous speed. Note: we remain in 1st stage machine control or fill stage. We now perform a velocity controlled pack at the reduced fill speed until the cavity is packed as determined by the post gate cavity pressure transducer.

The DIII processing technique has been proven with some mold/resin combinations to reduce the detrimental effects of resin viscosity variation. In other words, the DIII process is a process technique which helps to compensate for resin variation.

DIII processing can be difficult to set up the first time w/o experience or training. It is often worthwhile to first set mold up as DII process in order to identify the final cavity pressure desired and fill time desired; then convert process to DIII. The slow down to approx 10-15% of fill speed may need to be set a fraction sooner than the normal DII position for transfer. This slow down helps to prevent over pack caused by inertia and machine response time relative to the filled cavity.

---

[1] See also Injection Molding Reference Guide and/or Troubleshooting Guides for further discussion.

[2] DIII processing is a type of Decoupled Molding[sm] also taught by RJG, Inc in their Master Molder training w/ certification.

(Service Mark SM of RJG Associates, Inc)

Measurement System

In the earlier sections on variation, control charts and SQC, there were many sources of variation listed, including measurement variation. One key requirement for a good system is for the measurement variation to be much smaller than the part variation. The part variation does include measurement variation; thus, how do we know that measurement variation is smaller: this is done by performing an MSE – measurement system evaluation.

A rule of thumb is for the measuring equipment to contribute no more than 10% of the total variation. Sometimes, more (up to 30%) is accepted if the application is less critical and the cost to correct is high; thus, 10-30% is a judgement call based on Cpk or Z score risk, cost of defects, cost to reduce variation, cost to improve gage, etc.

In the pages discussing linear regression, outliers can have a major effect on the accuracy of the regression equation. Outliers have an effect on all data: the mean, std dev, control limits, etc; thus, outliers need to be identified and investigated.

Outliers could be real in that process fluctuations did occur during part collection, but in such events, there are often enough random process fluctuations whereby the data considers such outliers as normal variation and not a true outlier.

There are definitions as to what makes an outlier – how far from other data is it to be considered an outlier. One definition used by Minitab® is data that is more than 150% beyond the middle 50% of the data.

A measurement system evaluation is sometimes referred to as a gage R&R study (gage repeatability and reproducibility). The R&R refers to the gages ability to repeat same value on future measurements with same part, same person and same gage; the reproducibility refers to different operators (people) getting same value on same part and same gage.

Measurement error or variation is sometimes a significant part of the variation to be understood in a system. Often times the project engineering protocol requires a gage R&R study be performed. This is done as a safeguard toward understanding and developing an effective measurement and quality control system.

Unfortunately, some people think of the GR&R study as a another metric to be achieved. This results in the goal of an acceptable GR&R number ... sometimes skilled personnel learn just how to set up the test to yield a good number – that is the metric to be satisfied!

The real goal however should be to better understand the measurement system and correct any deficiencies before the same adversely effects process capability or qualification studies. If the measurement system is not optimized early on, then it can cause increased need for other corrective actions which may be needed to compensate for excessive variation. We already know there will be variation, that is inevitable – but if we cannot discern differences between process variation vs tooling variation vs measurement variation, it all gets attributed to the process and tooling.

Once we begin to talk about DOEs and other techniques such as the regression analysis to set SPC limits, then accurate results require accurate measurements. A general rule of thumb is: do not perform any DOEs or capability studies until the measurement system is tested and proven with acceptable results.

Measurement System Evaluation

Key terms to understand related to measurement system evaluations include the following:

Stability – consistency of measurements over time; same gage, same feature, same parts ... similar to repeatability except time related. In 6σ world, stability also requires the range chart to show control – all points below UCL after control charting the data.

Discrimination – good discrimination indicates the gage can discern differences in measurements; resolution of the gage, In 6σ world, the range chart would once again be reviewed, this time for an adequate number of measurement units below the UCL (five or more is typically good; look at different point levels). Poor discrimination may appear has poor stability whereby range chart is OOC.

Repeatability – aka precision; variation in measurements when same operator uses same gage, same feature, same parts; total variation from a central mean.

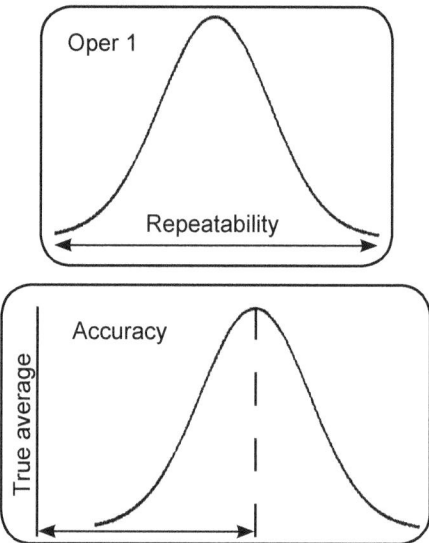

Accuracy – difference between observed value and true value.

Reproducibility – systematic variation or bias caused by the gage operator; variation inherent to the personnel; specifically variation in the average of measurements made by different operators, using the same gage, same feature and same parts.

Measurement System Evaluation (continued)

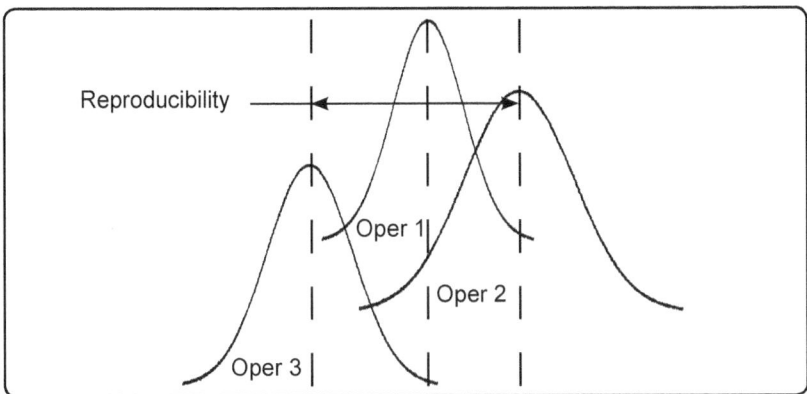

Bias Effects – the average of measurements are different by a similar amount; could be caused by an operator, a gage (e.g. one operator consistently gets readings approx. 0.0012 inches less or a certain gage measures and indicates a larger or smaller value ... operator bias and gage bias respectively).

NOTE:

$$\sigma_{part\ measurement} = \sqrt{\sigma^2_{part\ true\ value} + \sigma^2_{measurement\ error}}$$

The key point here is that the variation σ is comprised of two or more sources of variation, one of which is the measurement variation. Only the variance σ^2 is additive however; thus, in order to calculate the total variation, we must add the sum of the squares, then compute the square root to obtain the standard deviation. You will frequently see squares added in statistical operations because of this key point: only the variance (std dev squared) is additive. This equation can then be rewritten as follows:

$$\sigma_{part\ true\ value} = \sqrt{\sigma^2_{part\ measurement} - \sigma^2_{measurement\ error}}$$

We will use a similar calculation to separate repeatability from reproducibility in our gage R&R calculation ... see next two pages.

Gage R&R Study: EXCEL® Using Ranges

The following is a sample spreadsheet for a Gage R&R using the subgroup ranges to estimate standard deviation. This approach when there are repeats or reps AND multiple operators is known as the long method as it derives a repeatability number separate from reproducibility.

When performing gage R&R studies there are both long and short methods. The short method combines the repeatability and reproducibility together; thus, you cannot distinguish one from the other. The short method is often done with just two operators and five parts with just one measurement per part. The long method is often done with three operators, ten parts and three measurements on each part (2 operators and five parts could be chosen ... the key to achieving a separate repeatability number is to have repeat measurements on same parts by same operators).

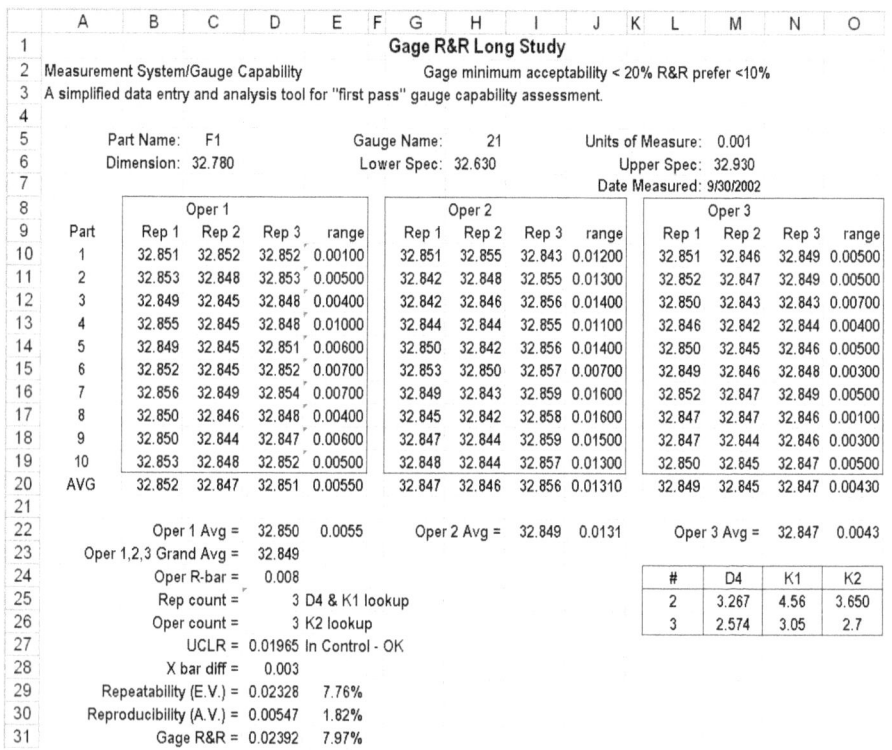

In order to comply with the newer standard of using 6 sigma instead of 5.15 sigma (99.7% of variation instead of 99%) ... You can change the constants above as follows:

K1 ... 4.56 becomes 5.319 & 3.05 becomes 3.544

K2 ... 3.650 becomes 4.252 & 2.7 becomes 3.146

#	D4	K1	K2
2	3.267	5.319	4.252
3	2.574	3.544	3.146

The results would change to: repeatability @ 9.02% & reproducibility @ 2.13% with total GR&R @ 9.27% for this example ... inflated values equal to a 6/5.15 multiplier!

Gage R&R Study: EXCEL® Using Ranges (cont.)

Cell Formulas from spreadsheet above are as follows:

D22=IF(ISERR(AVERAGE(B10:D19))," ",AVERAGE(B10:D19))

D23=AVERAGE(D22, I22, N22)

D24=AVERAGE(E22, J22, O22)

D25=COUNT(B10:D10)

D26=COUNT(D22, I22, N22)

D27=D24*VLOOKUP($D$25, $L$24:$O$26, 2)

E27=IF(MAX(E10:E19,J10:J19,O10:O19)>D27,"OOC: Check for root cause", "In Control-OK")

D28=MAX(D22, I22, N22)-MIN(D22, I22, N22)

D29=D24*VLOOKUP($D$25,$L$24:$O$26, 3)

E29=D29/(M6-H6)

D30=SQRT((D28*VLOOKUP($D$26,$L$24:$O$26,4))^2-(D29^2)/(D26*COUNT(B10:B19)))

E30=D30/(M6-H6)

D31=SQRT(D29^2+D30^2)

E31=SQRT(E29^2+E30^2)

Cell D30 rearranges the square root of the sum of the squares concept to separate repeatability from reproducibility.

$$\sigma_{reproducibility} = \sqrt{\sigma^2_{oper\ avgs} - \frac{\sigma^2_{repeatability}}{n}}$$

This concept will be seen more on next page, and thruout statistical math.

Gage R&R Study: EXCEL® Using Sigma

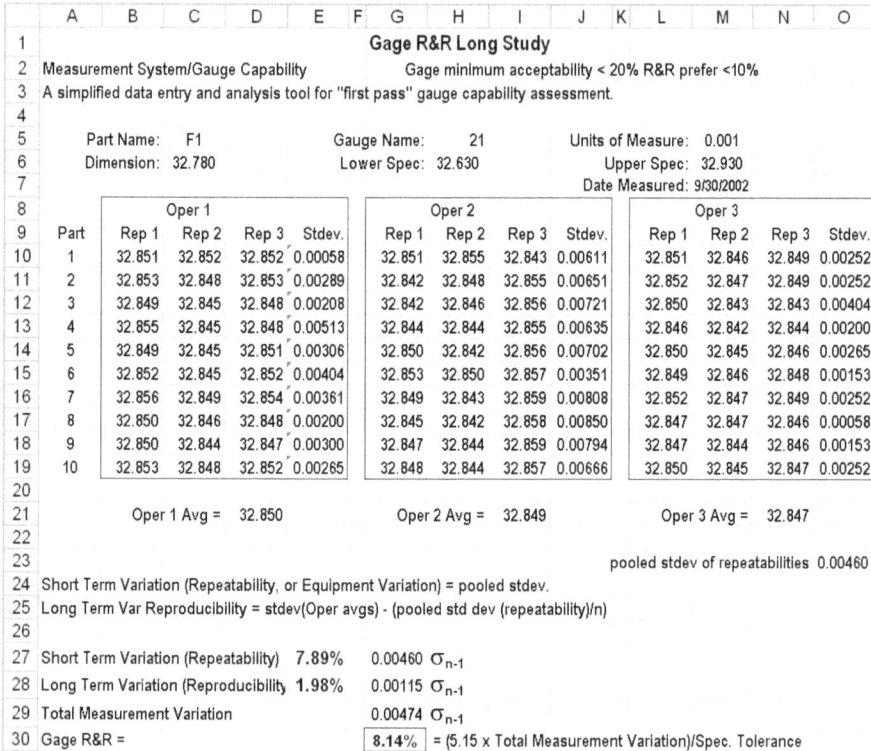

	A	B	C	D	E	F	G	H	I	J	K	L	M	N	O
1							Gage R&R Long Study								
2	Measurement System/Gauge Capability						Gage minimum acceptability < 20% R&R prefer <10%								
3	A simplified data entry and analysis tool for "first pass" gauge capability assessment.														
4															
5		Part Name:	F1				Gauge Name:	21				Units of Measure:	0.001		
6		Dimension:	32.780				Lower Spec:	32.630				Upper Spec:	32.930		
7												Date Measured:	9/30/2002		
8			Oper 1					Oper 2					Oper 3		
9	Part	Rep 1	Rep 2	Rep 3	Stdev.		Rep 1	Rep 2	Rep 3	Stdev.		Rep 1	Rep 2	Rep 3	Stdev.
10	1	32.851	32.852	32.852	0.00058		32.851	32.855	32.843	0.00611		32.851	32.846	32.849	0.00252
11	2	32.853	32.848	32.853	0.00289		32.842	32.848	32.855	0.00651		32.852	32.847	32.849	0.00252
12	3	32.849	32.845	32.848	0.00208		32.842	32.846	32.856	0.00721		32.850	32.843	32.843	0.00404
13	4	32.855	32.845	32.848	0.00513		32.844	32.844	32.855	0.00635		32.846	32.842	32.844	0.00200
14	5	32.849	32.845	32.851	0.00306		32.850	32.842	32.856	0.00702		32.850	32.845	32.846	0.00265
15	6	32.852	32.845	32.852	0.00404		32.853	32.850	32.857	0.00351		32.849	32.846	32.848	0.00153
16	7	32.856	32.849	32.854	0.00361		32.849	32.843	32.859	0.00808		32.852	32.847	32.849	0.00252
17	8	32.850	32.846	32.848	0.00200		32.845	32.842	32.858	0.00850		32.847	32.847	32.846	0.00058
18	9	32.850	32.844	32.847	0.00300		32.847	32.844	32.859	0.00794		32.847	32.844	32.846	0.00153
19	10	32.853	32.848	32.852	0.00265		32.848	32.844	32.857	0.00666		32.850	32.845	32.847	0.00252
20															
21		Oper 1 Avg =	32.850				Oper 2 Avg =	32.849				Oper 3 Avg =	32.847		
22															
23												pooled stdev of repeatabilities	0.00460		
24	Short Term Variation (Repeatability, or Equipment Variation) = pooled stdev.														
25	Long Term Var Reproducibility = stdev(Oper avgs) - (pooled std dev (repeatability)/n)														
26															
27	Short Term Variation (Repeatability)	7.89%			0.00460 σ_{n-1}										
28	Long Term Variation (Reproducibility)	1.98%			0.00115 σ_{n-1}										
29	Total Measurement Variation				0.00474 σ_{n-1}										
30	Gage R&R =				8.14% = (5.15 x Total Measurement Variation)/Spec. Tolerance										

NOTE: The industry typically now uses 6 instead of the 5.15 shown in line 30 above ... change accordingly. Note also that this will make your GR&R values approx 16.5 % higher (6/5.15). Change cells E27, E28 & G30 accordingly (6 in place of 5.15).

Gage R&R Study: EXCEL® Using Sigma (cont.)

G27= Repeatability = pooled std dev found as follows:
G27= SQRT(SUMSQ(E10:E19,J10:J19,O10:O19)/
     COUNT(E10:E19,J10:J19,O10:O19))
G28= Reproducibility =
     variance of Ops 1,2 & 3 avgs minus (pooled std dev ÷ n components thereof)

$$\sigma_{reproducibility} = \sqrt{\sigma^2_{oper\ avgs} - \frac{\sigma^2_{repeatability}}{n}}$$

computed as follows in EXCEL:

G28= SQRT(VAR(D21,I21,N21)-(O23^2)/(COUNT(B10:D19,G10:I19,L10:N19)/
COUNT(D21,I21,N21)))

G29= Total meas variation = sqr root of sum of squares for repeatability and reproducibility

$$\sigma_{measurement} = \sqrt{\sigma^2_{repeatability} + \sigma^2_{reproducibility}}$$

G30= 5.15*G29/(M6-H6) .... NOTE: The 5.15 multiplier has now been changed by the
industry to be 6 (5.15 based on 99% of variation vs 6 based on 99.7% of variation ...
change G30 accordingly (and E27 & E28 as shown on next two lines).

E27= 5.15*G27/(M$6-H$6)

E28= 5.15*G28/(M$6-H$6)

NOTE: The math operations using squared std dev (variance) in order to separate
different components of the variance.

The origin of the 5.15 is based on 2.575 std devs up and down from the grand average; this equates to 99% of a normal distribution.

Compare EXCEL® spreadsheet to MINITAB® results as follows:

	MINITAB®	EXCEL® (range)	EXCEL® (SD)
repeatability	0.00451		0.00460
reproducibility	0.00106		0.00115
total meas var	0.00463		0.00474
Gage R&R %	7.95%	7.97%	8.14%

Similar results are obtained; MINITAB® does offer some very good graphics (not
shown) to help draw conclusions.

Remember to always square the standard deviation if math operations are needed
such as addition, subtraction, averaging, etc. ... then convert back to standard deviation by taking square root. Standard deviation squared equal variance.

Gage R&R Study: MINITAB® Xbar&R ... Analysis

The data set at right is formatted as needed to be pasted into MINITAB® for analysis.

Parts 2-9 are hidden to save space; data is same as previous EXCEL spreadsheet data.

In MINITAB®, select: Stat, Quality Tools, Gage R&R (crossed), inside pop-up menu select: Xbar&R for results similar to EXCEL, (the ANOVA method is thought to do a better job of checking for oper x part interactions). Arrange data per example below.

Oper	Part	Rep	Width
1	1	1	32.851
1	1	2	32.852
1	1	3	32.852
1	10	1	32.853
1	10	2	32.848
1	10	3	32.852
2	1	1	32.851
2	1	2	32.855
2	1	3	32.843
2	9	3	32.859
2	10	1	32.848
2	10	2	32.844
2	10	3	32.857
3	1	1	32.851
3	1	2	32.846
3	1	3	32.849
3	10	1	32.850
3	10	2	32.845
3	10	3	32.847

```
Gage R&R for Width
                                    %Contribution
Source                  Variance    (of Variance)

Total Gage R&R          2.15E-05       93.13
  Repeatability         2.03E-05       88.24
  Reproducibility       1.13E-06        4.90
Part-to-Part            1.58E-06        6.87
Total Variation         2.30E-05      100.00


                        StdDev     Study Var   %Study Var   %Tolerance
Source                  (SD)       (5.15*SD)    (%SV)       (SV/Toler)

Total Gage R&R          4.63E-03   2.39E-02     96.51         7.95
  Repeatability         4.51E-03   2.32E-02     93.93         7.74
  Reproducibility       1.06E-03   5.47E-03     22.13         1.82
Part-to-Part            1.26E-03   6.48E-03     26.21         2.16
Total Variation         4.80E-03   2.47E-02    100.00         8.24
```

Compare 7.95 with 7.97 calculated in earlier pages using EXCEL spreadsheet.

Gage R&R Study: MINITAB® Xbar&R ... Chart

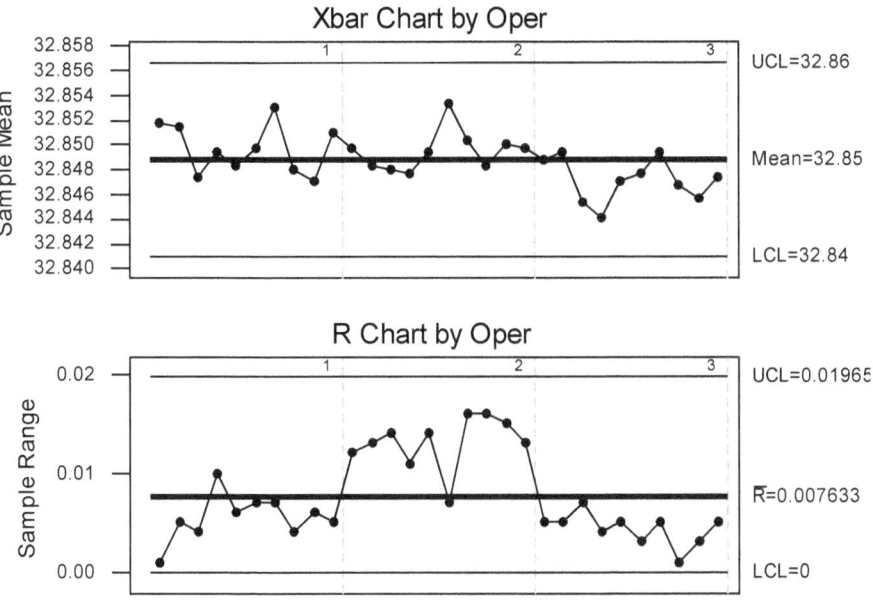

Note 1: Opers 1 & 2 are similar, but Oper 3 shows a bias downward ... training needed.

Note 2: There is adequate discrimination as evidenced by at least five layers of point values in range chart (look at points, not scale).

Note 3: There is adequate stability as evidenced by fact the range chart is in control.

Note 4: We don't really know if gage can check full spec range of part dimensions; good to have more dissimilar parts. We want to check for gage capability thruout the tolerance range (or more of it), Xbar chart should appear OOC if there is deliberate part variation to test gage... this is one time when the Xbar chart should be out of control limits (see next page). Sometimes, highly similar parts conditions the operator to look for a certain reading which can bias the results if operator is not fully objective; whereas, parts with variation help to deter expectations of dimensions.

Gage R&R Study: MINITAB® ANOVA

In this test, there was deliberate part variation to better test the gage. Now the repro-ducibility is at 23.98% (repeatability still OK). Oper 3 has significantly higher ranges than does Oper 1; Oper 3 also gets significantly higher highs and lower lows; thus, some technique refinement is needed to get reproducibility down to the desired <10% (hopefully combined R&R at <10%). The method used here is the <u>ANOVA method which is a more thorough analysis yielding higher numbers</u>; thus, take care in which method you use versus expectations. Note the Xbar&R analysis for same data.

```
Gage R&R
                                    %Contribution
Source                 VarComp      (of VarComp)
Total Gage R&R         0.0002082        12.47
  Repeatability        0.0000130         0.78
  Reproducibility      0.0001952        11.69
    Operator           0.0000368         2.20
    Operator*Part      0.0001584         9.49
Part-To-Part           0.0014610        87.53
Total Variation        0.0016692       100.00

                                 Study Var   %Study Var   %Tolerance
Source             StdDev (SD)   (5.15 * SD)    (%SV)      (SV/Toler)
Total Gage R&R      0.0144285      0.074307      35.32        24.77
  Repeatability     0.0036071      0.018577       8.83         6.19
  Reproducibility   0.0139703      0.071947      34.19        23.98
    Operator        0.0060625      0.031222      14.84        10.41
    Operator*Part   0.0125864      0.064820      30.81        21.61
Part-To-Part        0.0382230      0.196848      93.56        65.62
Total Variation     0.0408556      0.210406     100.00        70.14
```

Same data but with Xbar&R method:
% Tolerance (SV/Toler)
14.40
6.22
12.98
na
na
47.75
49.87

Gage R&R (ANOVA) for Part w/ Deliberate Variation - Test Gage

NOTE: The older analysis data above did use 5.15 SD instead of 6 as would be used today ... this would change the 24.77 to be 28.86.

Gage R&R Improvements

Other measurement issues with suggestions for improvement include the following:

1. Calibrate the gage with a standard.
2. Refer to operators by letter or number rather than names.
3. Do not have operator measure same part three times in a row to achieve the repeats; instead, have the operator measure a group of parts, then not remeasure same group until possibly other operators have measured. Frequently, this is not done and the operator measures same parts three times in a row which yields great repeatability numbers and consequently false confidence in the gage.
4. Do not let each operator see other operator's data during test due to risk for biasing measurements.
5. Measure parts in random order.
6. Record time, date and location of measurements AND any other pertinent facts regarding the surroundings ... such as temperature, etc. Sometimes, CMM equipment is located in non-optimum locations and surrounding equipment can be effecting the data.
7. Condition parts per normal inspection routine for Gage R&R: proper temperature and humidity and time (Nylon parts absorb moisture causing dimensional shift; some parts may require 48 hours to stabilize the shrinkage although many resins shrink 98% in first hour).

SOLUTIONS

Poor repeatability may be caused by gages with the following:

A. Springs or clamps which are too loose or too tight.
B. Gages such as calipers or micrometers whereby the measuring force is operator technique dependent (may be inconsistent even with same operator).
C. Part datums are not flat and locating proper with gage design.
D. Loose or poorly mounted indicators, clamps, etc.
E. Lack of recess clearance for parting lines, witness lines, or flash

Poor reproducibility may be caused by gages with the following:

F. Any of the above that a given operator might be skilled at compensating for; whereas, others do not.
G. Based on item F above; lack of training or standard gage use procedures.
H. Not randomizing the test and the gage setup becomes corrupted.
I. Different operators interpolate numbers not present in the readout device (correct by developing a procedure to address this).

Outlier Checks

Outliers are data values that do not seem to fit the pattern of normal variation.

Outliers are frequently caused by the following:

1. Data entry errors (if manually entered into analysis spreadsheet).
2. Measurement error from improper part placement into measurement fixture or fixture movement.
3. Other measurement error (loose components, etc).
4. Distorted or damaged part from handling (too many parts in too small of bag, molding damage from mold open or ejection, etc).
5. Data conversion errors if readout must be added to a standard.

The spreadsheets containing the data from qualification or process capability runs can be easily checked for outliers. Outliers can corrupt the data and resulting conclusions about variation present; sometimes to point of: creating added tool or process development to reduce variation to achieve a req'd Cpk, Ppk or Z value for customer requirements.

See next page for a sample data set with an outlier included. The outlier formulas are shown below table on next page. The outlier HI & LO threshold values are similar to the mean ± 3 std devs, but as can be seen, the outlier of 7.509 did not change the limits (same limits in column G vs F after outlier was rechecked to be 7.504).

Conditional formatting can be used to highlight outliers.

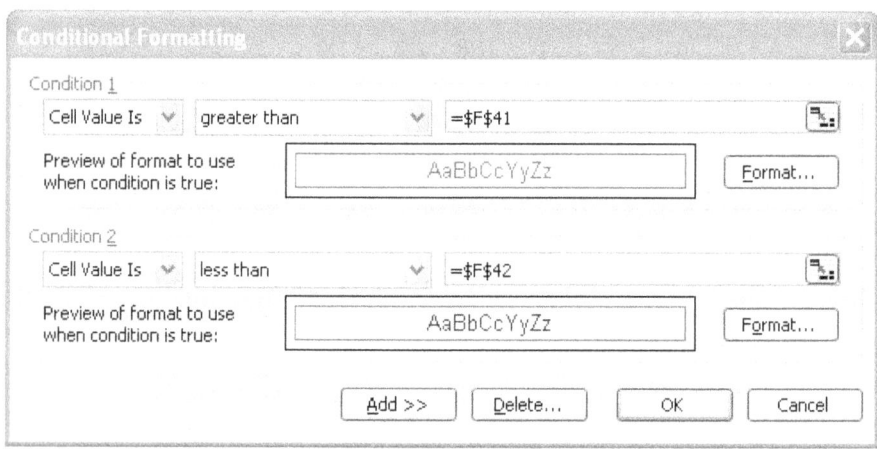

NOTE: The top preview of format is red text and bottom is blue text ... not visible in this gray scale graphic ... formatting is user specified with format box on right. This is a commonly used, but useful formatting tool in Microsoft EXCEL®.

Outlier Checks: Spreadsheet

	A	B	C	D	E	F	G
1					nom	7.500	7.500
2					USL	7.510	7.510
3					LSL	7.490	7.490
4	sample	Lot	Mold	Insp	Gage Rep	F1	F1
5	1	106	A	George	1	7.5030	7.5030
6	2	106	A	George	1	7.5020	7.5020
7	3	106	A	George	1	7.5020	7.5020
8	4	106	A	George	1	7.5010	7.5010
9	5	106	A	George	1	7.5020	7.5020
10	6	106	A	George	1	7.5040	7.5040
11	7	106	A	George	1	7.5010	7.5010
12	8	106	A	George	1	7.5020	7.5020
13	9	106	A	George	1	7.5040	7.5040
14	10	106	A	George	1	7.5050	7.5050
15	11	106	A	George	1	7.5040	7.5040
16	12	106	A	George	1	7.5030	7.5030
17	13	106	A	George	1	7.5040	7.5040
18	14	106	A	George	1	7.5040	7.5040
19	15	106	A	George	1	7.5030	7.5030
20	16	106	A	George	1	7.5040	7.5040
21	17	106	A	George	1	7.5020	7.5020
22	18	106	A	George	1	7.5030	7.5030
23	19	106	A	George	1	7.5040	7.5040
24	20	106	A	George	1	7.5030	7.5030
25	21	106	A	George	1	7.5090	7.5040
26	22	106	A	George	1	7.5040	7.5040
27	23	106	A	George	1	7.5020	7.5020
28	24	106	A	George	1	7.5020	7.5020
29	25	106	A	George	1	7.5040	7.5040
30	26	106	A	George	1	7.5040	7.5040
31	27	106	A	George	1	7.5020	7.5020
32	28	106	A	George	1	7.5030	7.5030
33	29	106	A	George	1	7.5040	7.5040
34	30	106	A	George	1	7.5040	7.5040
35					average	7.5033	7.5031
36					std dev	0.00151	0.00106
37					max	7.5090	7.5050
38					min	7.5010	7.5010
39					centering	67.3%	69.0%
40					count	30	30
41					outlier HI	7.5070	7.5070
42					outlier LO	7.4990	7.4990
43					Pp	2.21	3.14
44					Ppk	1.49	2.17
45					Z_{LSL}	8.80	12.34
46					Z_{USL}	4.47	6.50
47					PPM	4	0

Cell F25 exceeds the outlier HIGH or LARGE threshold found in cell F41; thus, it is shaded ... the formulas for outlier threshold values are shown below:

Outlier HIGH or LARGE threshold = Cell F41=
LARGE(F$5:F$34,F$40/4)+1.5*(LARGE(F$5:F$34,F$40/4)-SMALL(F$5:F$34,F$40/4))

Outlier LOW or SMALL threshold = Cell F42=
SMALL(F$5:F$34,F$40/4)-1.5*(LARGE(F$5:F$34,F$40/4)-SMALL(F$5:F$34,F$40/4))

Components of Variance (COV)

Review the following three groups of data for variation ... discussion continued on next pages.

	A	B	C	D	E	F	G	H	I	J
1		FIRST MOLD SAMPLING						RUN 2		PROD
2										7.500
3	This mold was built oversize (intentionally);									7.510
4	thus, steel was resized by -0.040 inches.									7.490
5										F1
6										7.5040
7										7.5030
8					nom	7.500		7.500		7.5090
9					USL	7.510		7.510		7.5070
10					LSL	7.490		7.490		7.5120
11	sample	Lot	Mold	Insp	Gage Rep	F1		F1		7.5130
12	1	106	A	George	1	7.5430		7.5030		7.5140
13	2	106	A	George	1	7.5420		7.5020		7.5160
14	3	106	A	George	1	7.5420		7.5020		7.5000
15	4	106	A	George	1	7.5410		7.5010		7.5010
16	5	106	A	George	1	7.5420		7.5020		7.5040
17	6	106	A	George	1	7.5440		7.5040		7.5070
18	7	106	A	George	1	7.5410		7.5010		7.5040
19	8	106	A	George	1	7.5420		7.5020		7.5030
20	9	106	A	George	1	7.5440		7.5040		7.5080
21	10	106	A	George	1	7.5450		7.5050		7.5080
22	11	106	A	George	1	7.5440		7.5040		7.5020
23	12	106	A	George	1	7.5430		7.5030		7.5010
24	13	106	A	George	1	7.5440		7.5040		7.5060
25	14	106	A	George	1	7.5440		7.5040		7.5070
26	15	106	A	George	1	7.5430		7.5030		7.5010
27	16	106	A	George	1	7.5440		7.5040		7.5000
28	17	106	A	George	1	7.5420		7.5020		7.5040
29	18	106	A	George	1	7.5430		7.5030		7.5050
30	19	106	A	George	1	7.5440		7.5040		7.4920
31	20	106	A	George	1	7.5430		7.5030		7.4930
32	21	106	A	George	1	7.5440		7.5040		7.5020
33	22	106	A	George	1	7.5440		7.5040		7.5050
34	23	106	A	George	1	7.5420		7.5020		7.4960
35	24	106	A	George	1	7.5420		7.5020		7.4980
36	25	106	A	George	1	7.5440		7.5040		7.5020
37	26	106	A	George	1	7.5440		7.5040		7.5060
38	27	106	A	George	1	7.5420		7.5020		7.4900
39	28	106	A	George	1	7.5430		7.5030		7.4910
40	29	106	A	George	1	7.5440		7.5040		7.5020
41	30	106	A	George	1	7.5440		7.5040		7.5040
42					average	7.5431		7.5031		7.5033
43					std dev	0.00106		0.00106		0.00603
44					max	7.5450		7.5050		7.5160
45					min	7.5410		7.5010		7.4900
46					centering	-331.0%		69.0%		66.7%
47					count	30		30		36
48					outlier HI	7.547		7.507		7.516
49					outlier LO	7.539		7.499		7.492
50					Pp	3.14		3.14		0.55
51					Ppk	-10.39		2.17		0.37
52					Z_{LSL}	50.01		12.34		2.21
53					Z_{USL}	-31.17		6.50		1.10
54					PPM	1000000		0		148135

From the table above, we have gone from Run #2 with 0 PPM defective to PROD with 148,135 PPM defects; we need to drill deeper to better understand the PROD data.

How can the total variation be separated into basic components for source identification, prioritization and improvement. There are techniques whereby the variance is separated into separate components of variance.

Components of Variance (COV) ... continued

Consider the data progression shown whereby Run #2 after steel resizing; shows good Ppk values, but the prod data has a poor composite Ppk, similar mean, but high std dev ... the prod data includes three molds, three resin lots and two inspectors.

During tooling and process development, there are experimentation techniques known as DOE (Design of Experiments) which can be utilized to better understand factor effects on responses. COV analysis is sometimes more applicable when trying to drill down into historical data to draw conclusions from work already done such as data from previous production. The COV analysis identifies the variation in the process; once identified and quantified: continuous improvement priorities can be established.

It should be noted however that the historical data must be accurate. Be wary of production data whereby the lot numbers were not changed timely with the actual lot changes: this can corrupt the conclusions. For this reason, good record keeping discipline must be emphasized and maintained within an organization.

Nested data analysis such as the Fully Nested ANOVA may be applicable to draw conclusions from historical data.

The standard Xbar/R chart is a simplified example of a nested analysis. There is analysis of variation for within subgroups and between subgroups. When reviewing a Xbar/R chart; we typically draw the following conclusions: if the range chart is in control, but Xbar chart is out of control – we have unexplained special caused variation, and we need to look elsewhere in the hierarchical relationships.

The typical COV analysis performs different subgrouping strategies on the same data to yield different Rbars. The different Rbars ÷ d_2 yield estimates of the different variances present. The typical Gage R&R study performs this same COV type analysis to separate components of variance.

The components of variance from data on previous page include the following listed sources. We must remember that only the variances are additive (std dev squared). We get a relationship as follows:

$$\sigma^2_{total} = \sigma^2_{lot} + \sigma^2_{mold} + \sigma^2_{inspector} + \sigma^2_{measurement}$$

This variance may be estimated; such as:
$$\frac{\bar{R}}{d_2} = \hat{\sigma} \quad and \quad \left(\frac{\bar{R}}{d_2}\right)^2 = \hat{\sigma}^2$$

There could be many more sources of variation studied such as cavity to cavity, specific gages (if more than one), specific molding presses (if more than one), shift, time, auxiliary equipment (if tracked and known), etc. Large amounts of variation in one of these levels may identify the need to look closer. A detailed process map can help to identify all the possible sources of variation: it should list all the inputs (and outputs) to the process. Inputs which are quantitative can be evaluated for effect.

The next several pages will further discuss COV analysis. If certain inputs contain large amounts of variation that cannot be further separated, then DOEs may be required. DOEs will be discussed in greater detail in this book. Sometimes DOEs are done to identify factors (inputs) needed to effect the response, but DOEs can also use variance as the response; thus, ID inputs needed for lowest output variance.

Components of Variance: Data Set

The middle data was provided by the quality department. The data was merged into the standard Engineering Analysis Form so the total risk could be better understood and plan the corrective actions needed.

From prod QC Dept ...data inserted into Engrg Analysis Form

				nom	7.500
				USL	7.510
				LSL	7.490
Sample	Lot	Mold	Insp	Gage Rep	F1
1	123	A	George	1	7.5040
2	123	A	George	2	7.5030
3	123	A	Ann	1	7.5090
4	123	A	Ann	2	7.5070
5	123	B	George	1	7.5120
6	123	B	George	2	7.5130
7	123	B	Ann	1	7.5140
8	123	B	Ann	2	7.5160
9	123	C	George	1	7.5000
10	123	C	George	2	7.5010
11	123	C	Ann	1	7.5040
12	123	C	Ann	2	7.5070
13	124	A	George	1	7.5040
14	124	A	George	2	7.5030
15	124	A	Ann	1	7.5080
16	124	A	Ann	2	7.5080
17	124	B	George	1	7.5020
18	124	B	George	2	7.5010
19	124	B	Ann	1	7.5060
20	124	B	Ann	2	7.5070
21	124	C	George	1	7.5010
22	124	C	George	2	7.5000
23	124	C	Ann	1	7.5040
24	124	C	Ann	2	7.5050
25	125	A	George	1	7.4920
26	125	A	George	2	7.4930
27	125	A	Ann	1	7.5020
28	125	A	Ann	2	7.5050
29	125	B	George	1	7.4960
30	125	B	George	2	7.4980
31	125	B	Ann	1	7.5020
32	125	B	Ann	2	7.5060
33	125	C	George	1	7.4900
34	125	C	George	2	7.4910
35	125	C	Ann	1	7.5020
36	125	C	Ann	2	7.5040
				average	7.5033
				std dev	0.00603
				max	7.5160
				min	7.4900
				centering	66.7%
				count	36
				Pp	0.55
				Ppk	0.37
				Z_{LSL}	2.21
				Z_{USL}	1.10
				PPM	148135

COV: MINITAB® Nested ANOVA Data Analysis

If we rotate the data seen on previous page clockwise, we can begin to see data nested in the following hierarchical relationships:

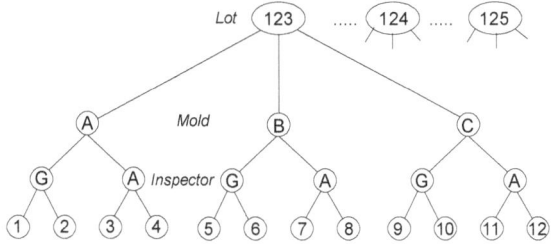

The measurement repeatability (2x) is nested within each inspector (2x), which is nested within each mold (3x) which is nested within the different lot (3x). We need to separate the components of variance so we know the contribution from each source of the total (what percent is each individual component).

This can be done in Microsoft® EXCEL in a similar manner as separating repeatability from reproducibility: subtract variances divided by n value making up that specific variance ... this technique is <u>extremely difficult</u> in any spreadsheet software – see next page.

As stated on next pages, the best and simplest choice is to use software such as MINITAB®. Here we just select: Stat, ANOVA, Fully Nested ANOVA (list all the factors in the factors box not shown ... in Minitab®). The data output shown below indicates % output for each variance component. The inspector gage-use training should be the focus of our first efforts: MSE (measurement system evaluation) is clearly needed. The resin lot variation should also be investigated. The repeatability will be reviewed in the MSE. After the aforementioned focus, the molds can be reviewed for improvement opportunities, but this data clearly shows two other items contributing most of the variation.

This type analysis clearly identifies the priority needed to each COV – component of variance. This data is historical data; thus, no experiments required!

```
Nested ANOVA: F1 versus Lot, Mold, Insp, Gage Rep
Analysis of Variance for F1
```

Source	DF	SS	MS	F	P
Lot	2	0.0005	0.0003	5.054	0.052
Mold	6	0.0003	0.0000	1.024	0.467
Insp	9	0.0004	0.0000	29.267	0.000
Gage Rep	18	0.0000	0.0000		
Total	35	0.0013			

```
Variance Components
```

Source	Var Comp.	% of Total	StDev
Lot	0.000	39.82	0.004
Mold	0.000	0.70	0.001
Insp	0.000	55.55	0.005
Gage Rep	0.000	3.93	0.001
Total	0.000		0.007

NOTE: The historical data must be accurate for good results.

COV: Spreadsheet Analysis

This analysis was done in EXCEL® and can be difficult to set up and interpret without error:

	A	B	C	D	E	F	G	H	I	J	K	L	M
1						rep avg	rep range	insp avg	insp range	mold avg	mold range	lot avg	lot range
2	Lot	Mold	Insp	Rep	F2								
3	123	A	George	1	7.504								
4	123	A	George	2	7.503	7.504	0.001	7.504					
5	123	A	Ann	1	7.509								
6	123	A	Ann	2	7.507	7.508	0.002	7.508	0.005	7.506			
7	123	B	George	1	7.512								
8	123	B	George	2	7.513	7.513	0.001	7.513					
9	123	B	Ann	1	7.514								
10	123	B	Ann	2	7.516	7.515	0.002	7.515	0.003	7.514			
11	123	C	George	1	7.500								
12	123	C	George	2	7.501	7.501	0.001	7.501					
13	123	C	Ann	1	7.504								
14	123	C	Ann	2	7.507	7.506	0.003	7.506	0.005	7.503	0.011	7.508	
15	124	A	George	1	7.504								
16	124	A	George	2	7.503	7.504	0.001	7.504					
17	124	A	Ann	1	7.508								
18	124	A	Ann	2	7.508	7.508	0.000	7.508	0.005	7.506			
19	124	B	George	1	7.502								
20	124	B	George	2	7.501	7.502	0.001	7.502					
21	124	B	Ann	1	7.506								
22	124	B	Ann	2	7.507	7.507	0.001	7.507	0.005	7.504			
23	124	C	George	1	7.501								
24	124	C	George	2	7.500	7.501	0.001	7.501					
25	124	C	Ann	1	7.504								
26	124	C	Ann	2	7.505	7.505	0.001	7.505	0.004	7.503	0.003	7.504	0.003
27	125	A	George	1	7.492								
28	125	A	George	2	7.493	7.493	0.001	7.493					
29	125	A	Ann	1	7.502								
30	125	A	Ann	2	7.505	7.504	0.003	7.504	0.011	7.498			
31	125	B	George	1	7.496								
32	125	B	George	2	7.498	7.497	0.002	7.497					
33	125	B	Ann	1	7.502								
34	125	B	Ann	2	7.506	7.504	0.004	7.504	0.007	7.501			
35	125	C	George	1	7.490								
36	125	C	George	2	7.491	7.491	0.001	7.491					
37	125	C	Ann	1	7.502								
38	125	C	Ann	2	7.504	7.503	0.002	7.503	0.013	7.497	0.004	7.498	0.006
39	Xbar or Rbar >>>>>>>>>>>>>>>>>>>>>					7.503	0.00156	7.503	0.00622	7.503	0.00592	7.503	0.00454
40	A2 or (3/d2 = 2.66 for MR) or d4 for range limits					1.88	3.267	1.88	3.267	1.023	2.574	2.66	3.267
41	LCL					7.500	0.000	7.492	0.000	7.497	0.000	7.491	0.000
42	UCL					7.506	0.005	7.515	0.020	7.509	0.015	7.515	0.015
43						OUT	IN	OUT	IN	OUT	IN	IN	IN
44	$(\sigma_r)^2$	4.69%	rep			1.90E-06							
45	d2					1.128							
46	$(\sigma_I)^2$	72.75%	insp					2.95E-05					
47	d2							1.128					
48	$(\sigma_M)^2$	-7.41%	mold							-3.00E-06			
49	d2									1.693			
50	$(\sigma_L)^2$	29.96%	lot									1.21E-05	
51	d2											1.128	
52	$(\sigma_T)^2$	100.00% >>>>>>>>>>>>>>>>>>>				4.05E-05 <<<<<<<<<<<<< =SUM ALL VARIANCES							
53	STD DEV (E3:E38) = 0.00603 <VS>					0.00637 <<<<<<<<<<<<< =STD DEV OF VARIANCE SUMS							

NOTE: Interpret negative numbers as zero.

$$\hat{\sigma}_L^2 = \left(\frac{\bar{R}}{d_2}\right)^2 - \left(\frac{\hat{\sigma}_M^2}{3}\right) - \left(\frac{\hat{\sigma}_I^2}{3 \times 2}\right) - \left(\frac{\hat{\sigma}_R^2}{3 \times 2 \times 2}\right)$$

COV: Spreadsheet Formulas

The best solution is to use software such as MINITAB® to perform this type COV analysis. This example is provided more to show the formula at bottom of previous page indicating methodology of subtracting out variances.

$$\hat{\sigma}_L^2 = \left(\frac{\bar{R}}{d_2}\right)^2 - \left(\frac{\hat{\sigma}_M^2}{3}\right) - \left(\frac{\hat{\sigma}_I^2}{3 \times 2}\right) - \left(\frac{\hat{\sigma}_R^2}{3 \times 2 \times 2}\right)$$

Selected formulas for spreadsheet on previous page are listed below:

cell L50=(M39/L51)^2-(J48/3)-(H46/6)-(F44/12) ... same as formula above!

cell H46=(I39/H47)^2-(F44/2)

cell J48=(K39/J49)^2-(H46/2)-(F44/(2*2))

cell F52=F44+H46+J48+L50

Determine divisor n values by looking at nesting chart (hierarchical order on earlier page); determine how many separate values of each contribution layer.

NOTE:
R variance divisor of 3x2x2 results from 3 molds x 2 inspectors x 2 repeats
I variance divisor of 3x2 results from 3 molds x 2 inspectors
M variance divisor of 3 results from 3 molds

Note once again ... the best solution is to use software such as MINITAB® to perform this type COV analysis. This technique is too powerful for drilling down into historical data to NOT perform because of very difficult math; thus, use the software designed for this technique. COV analysis not only yields valuable insight to "components of variance" or sources or variation ... but the best part is: the data typically is already there; no DOEs are required. This is data mining at it's finest!

COV: MINITAB®

There are three methods/choices in MINITAB® for evaluating components of variance:

1. Balanced ANOVA
2. General Linear Model
3. Fully Nested ANOVA

The first two will test for crossed terms; whereas, the third (Fully Nested) does not. A crossed term is basically an interaction term. If measurements are made by both inspectors on both molds, then these two factors are crossed (possible interaction). The model should be set up accordingly which cannot be done in the Fully Nested ANOVA.

Some prefer to ask themselves: is it logical or is it probable that an interaction is present between the inspector and mold. If not, the Fully Nested ANOVA is easier to setup for analysis (note however: it is sometimes hard to predict if an interaction might be present). An interaction could easily exist between resin and mold (poor cooling vs nucleation rate) or mold and cavity if cavity separation was listed in nesting structure (imbalance of fill issues, gate sizes, etc). Crossed terms do take some additional skill in setting up properly. There are some additional data outputs which can be useful such as the means computed for each component. Various additional interactions can be tested, but the conclusion for this data set is still: insp is 1st priority followed by lot. It does indicate the mold to also be significant.

Run as follows in Minitab®: Stat, ANOVA, Balanced ANOVA ... Menus below, data analysis results on next page.

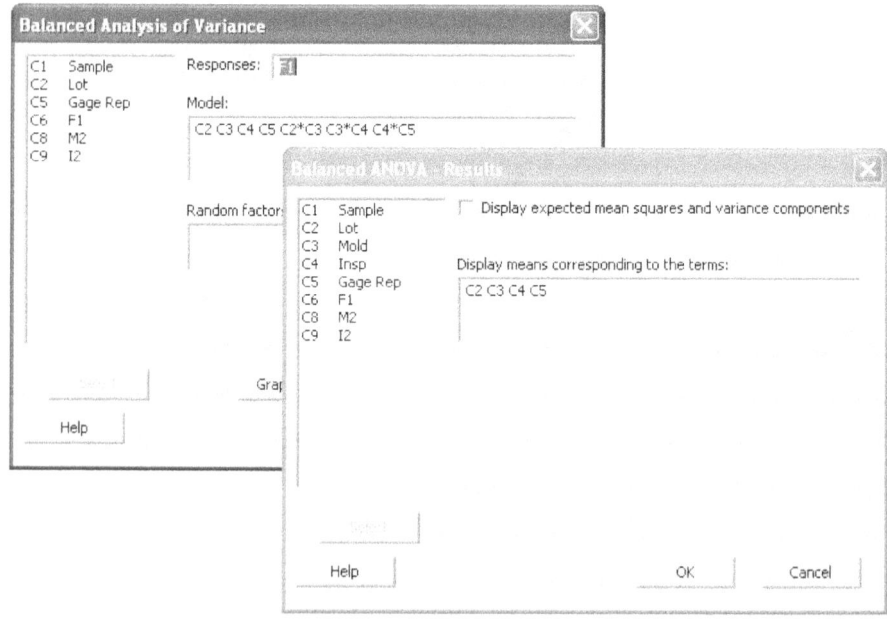

COV: MINITAB® ... Balanced ANOVA

(Balanced) ANOVA: F1 versus Lot, Mold, Insp, Gage Rep

Factor	Type	Levels	Values
Lot	fixed	3	123, 124, 125
Mold	fixed	3	A, B, C
Insp	fixed	2	Ann, George
Gage Rep	fixed	2	1, 2

Analysis of Variance for F1

Source	DF	SS	MS	F	P
Lot	2	0.000505167	0.000252583	55.35	0.000
Mold	2	0.000171167	0.000085583	18.76	0.000
Insp	1	0.000348444	0.000348444	76.36	0.000
Gage Rep	1	0.000007111	0.000007111	1.56	0.225
Lot*Mold	4	0.000128667	0.000032167	7.05	0.001
Mold*Insp	2	0.000009056	0.000004528	0.99	0.387
Insp*Gage Rep	1	0.000004000	0.000004000	0.88	0.359
Error	22	0.000100389	0.000004563		
Total	35	0.001274000			

S = 0.00213615   R-Sq = 92.12%   R-Sq(adj) = 87.46%

Means

Lot	N	F1
123	12	7.5075
124	12	7.5041
125	12	7.4984

The analysis of means indicates "lot" to have larger mean differences, BUT you need to trust the ANOVA table above which indicates the inspector to be more statistically significant.

Mold	N	F1
A	12	7.5032
B	12	7.5061
C	12	7.5008

The analysis of means indicates mold "B" might need to be resized.

Insp	N	F1
Ann	18	7.5064
George	18	7.5002

The analysis of means indicates "lot" to have larger mean differences and sizeable mold differences, BUT you need to trust the ANOVA table above which indicates the inspector to be more statistically significant.

Gage Rep	N	F1
1	18	7.5029
2	18	7.5038

COV Data & Boxplots

Boxplots can be displayed by MINITAB® in an effort to quickly discern graphical differences: (Select: Graph, Boxplot, One Y, With Groups, OK, ... graph variables = F1 ... categorical variables = Lot ... then do same for Mold, etc.).

1. The top boxplot for lot clearly shows some differences for the different lots.
2. The molds look to be more similar except much more variation with mold B.
3. The inspectors have different averages, but also have different patterns of variability...they may be able to learn from each other or at least become more similar in performance.

There are a few outliers (*) in this data which emphasizes the need for metrology improvements. See also earlier page on outliers in spreadsheet analysis.

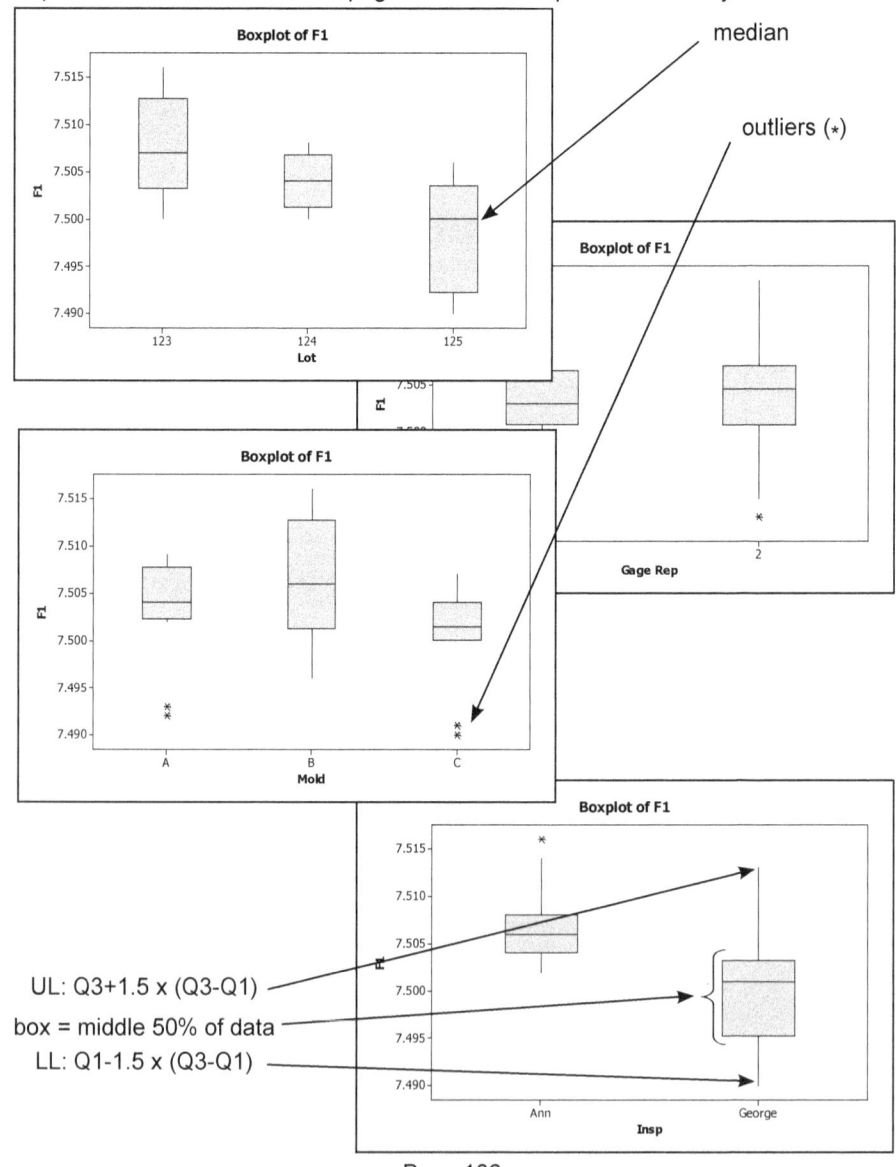

Blank Page

Design Of Experiments (DOE)

What is a DOE?

A systematic method of making controlled process changes to identify main effects (and interactions thereof) for purposes of altering (improving) that process. This is done using the predictive (regression) equation outputs from a DOE analysis. Linear regression will also output such predictive equations, but the DOE looks at many more variables simultaneously and identifies any significant interactions among factors.

Since DOEs look at interactions as well, the experiment can become quite large in terms of the number of unique process combinations to be run. The total number of combinations is determined by taking the number of levels (typically 2) to an exponential value equal to the number of factors.

For example:

$2^3 = 8$ ... (2 levels^3 ^{factors} = 8 combinations ... if all combos need to be analyzed).
4 factors = 16 combos, 5 factors = 32 combos, etc. See table below for 3 factor, 2 level experiment.

Process Combination	Stop Pin A	Angle B	Tension C	AB	AC	BC	ABC
		Factors & Levels		**INTERACTIONS**			
1	-1	-1	-1	1	1	1	-1
2	1	-1	-1	-1	-1	1	1
3	-1	1	-1	-1	1	-1	1
4	1	1	-1	1	-1	-1	-1
5	-1	-1	1	1	-1	-1	1
6	1	-1	1	-1	1	-1	-1
7	-1	1	1	-1	-1	1	-1
8	1	1	1	1	1	1	1

The table above shows a matrix which yields a menu of processes to perform. We will need to set process conditions for each variable above: Stop Pin, Angle & Tension, there will be a HI condition and a LO condition – real values or conditions, not just 1 or -1. The menu guides us to run 8 process combinations whereby we set the process with the conditions shown. For example: Process 1 has all three variables at the LO level (-1) ... NOTE: The 1s & -1s will be replaced with actual process conditions. The 1s & -1s merely mean HI & LO, but the signs of + & - are used in later math. This later math results in the table values being multiplied by one or more +1 or -1 ... such as AB above in process combo 1 displays a +1 ... this comes from A @ -1 x B @ -1 to yield +1. ABC in process combo 1 shows a -1 ... from -1 x -1 x -1 = -1.

DOE (continued)

The table below is a table of average differences (analysis of means) and, is set up in Microsoft® EXCEL. The data is for a basic statapult[1] experiment. Even with simple spreadsheet software, it can be seen that the angle has greatest net effect on the response (distance), and tension has the next greatest effect. The interactions are low.

Y							
AVG	Stop Pin	Angle	Tension		INTERACTIONS		
Response	A	B	C	AB	AC	BC	ABC
58.3	-58.3	-58.3	-58.3	58.3	58.3	58.3	-58.3
59.5	59.5	-59.5	-59.5	-59.5	-59.5	59.5	59.5
90.8	-90.8	90.8	-90.8	-90.8	90.8	-90.8	90.8
104.9	104.9	104.9	-104.9	104.9	-104.9	-104.9	-104.9
74.0	-74.0	-74.0	74.0	74.0	-74.0	-74.0	74.0
90.9	90.9	-90.9	90.9	-90.9	90.9	-90.9	-90.9
134.6	-134.6	134.6	134.6	-134.6	-134.6	134.6	-134.6
141.8	141.8	141.8	141.8	141.8	141.8	141.8	141.8
Effect >>	9.8	47.3	32.0	0.8	2.2	8.4	-5.7

The effects above are computed in last line and is calculated by taking the difference between the average of values observed at HI minus average of values observed at LO. In example above, we calculated the 9.8 (average effect of A which is Stop Pin), located in column N (not labeled) in spreadsheet as follows:

=SUMIF(N8:N15,">0")/((COUNT(N8:N15)/2))- ABS(SUMIF(N8:N15,"<0")/((COUNT(N8:N15)/2)))

We could make this formula different to compute this and get same answer ... this spreadsheet constructed as such to permit copying cell to other effects and/or interactions. A simple sum total of column including positive and negative values would yield different values, but their values relative to each other would be accurate as far as relative strength.

We do not know if these results are statistically significant which is why we typically use software using the ANOVA method (analysis of variance). The ANOVA can test for statistical significance of the factors. ANOVA fundamentals and how it tests for statistical significance will be discussed in later pages.

We can see; however, that the interactions are relatively weak (near zero) and the angle of launch has greatest effect followed by the spring tension, followed by stop pin as a relative order of the strength of process condition effects. The main effects above were all positive, but if there was a negative, it just means an indirect relationships (i.e. higher process conditions yield lower values results and vise versa).

---

[1] Statapult is a small catapult commonly used in statistical training; search statapult on internet

DOE (continued)

Why Perform DOEs?
The purpose of a DOE is typically one of the following (we may need to perform all three for maximum effectiveness):
1. <u>Screening DOEs</u> to identify most important process factors. Frequently there is not enough profound process knowledge to know which process factors are most important; those who profess to know are sometimes mistaken.
2. <u>Characterize</u> the process: this includes checking a mold's (or molded product's) sensitivity to process changes - testing the inference space - how much can a given response be affected.
3. <u>Optimize</u> the process: develop predictive equations for purposes of process correcting a dimensional deficiency. Move dimensions closer to nominal to reduce risk of defects. Reduce variation to minimize risk.

Responses
A "response" is the dimension or physical property, variance, etc. that is to be measured or quantified as a result of each process combination (aka treatment combination or factor combination). Responses must be variables type data since the analysis is mathematical (instead of attribute data).

Attribute data is simply pass/fail, good/bad or go/no-go types of data. Attribute data such as good or bad appearance with regards to sinks (etc) can be converted into numbers (e.g. scale of 1-5) for inclusion into the experiment's data analysis. There are techniques of how best to convert such attribute data to numbers (see also Kappa studies and Intraclass correlation – ICC – not covered in this book).

Preparation
Before running a DOE, it is necessary to discuss the DOE for purposes of selecting the responses, the factors and the levels. This is typically best accomplished by discussing the DOE with those most familiar with the mold, machine, resin and the process. It is also advisable to experiment with selected parameters to verify that all combinations of factors and levels are doable without resulting in short shots, flash or otherwise unacceptable products. Experienced processors can usually identify the combinations which will interact to reflect the most extreme process condition for preliminary tests of feasibility. It should be noted however, that the actual conclusions as to which variables have most effect on responses are the result of statistical analysis on the DOE data. Prior to running the DOE, you should also have already:
- Established the balance of fill as being acceptable.
- Determined the needed overall cycle time (unless in the DOE).
- Determined the pack & hold pressure <u>limitations</u>.
- Determined the gate seal time (select an overall injection forward time).

Factors and Levels
Factors to be investigated must be chosen. They can be any of the controllable inputs to the process. Skill is required in factor selection; even more skill is required in level specification. If the levels are too narrow, then a truly significant factor may not appear to be significant; conversely if the levels are too broad, then you can force insignificant factors into being significant. The nozzle orifice size of a molding press is not typically among the most important factors in a molding press – unless sized poorly. If included as a factor in a DOE whereby one size was 0.045 inches in diameter and the other size was 0.250 inches (assumes sprue orifice is > 0.250 inches); the nozzle may show up as significant because the 0.045 is excessively small.

DOE (continued)

There are many variables in injection molding and nearly all of them could be forced into significance if poorly set, but there is engineering logic to be applied for many factors (e.g.):
1. Supplier prescribed process temperature ranges.
2. Nozzles as large as possible, but smaller than sprue orifice.
3. Pack psi sufficient for full parts w/o sink, but avoid flash.
4. Pack times to achieve a plateau on the gate seal study.
5. Fill times w/o flash & w/o diesel burns, but not so slow to compromise fill viscosity per fill time study.

Prior to running the experiments, the process parameters are reviewed and selected based on process knowledge. We often have a feel for which parameters are key to influencing certain responses. In the injection molding industry, there are four process factors which are useful toward evaluating product sensitivity to process variation and resulting product performance (a screening DOE may identify others that are important for your application).
*   melt temperature (± approx 15°- 25° F from nominal)
*   injection speed or fill time (± approx 30% from nominal, if possible)
*   packing psi (± approx 100 psi hydraulic from nominal, if possible)
*   mold temperature (± approx 10°- 20° F from nominal)

Note: If mold temp differentials are tested; limit max diff to approx 30° F.

The selection of these parameters assumes reasonable selection of packing time, and the many other process variables which are easily set, but can have adverse effects if poorly set (per above discussion). The machine must also be stable and exhibit an "in control" pattern of operation. We may select only 3 or 4 of these factors, and we may decide on two mold temps to test mold temperature differentials. We should always include packing pressure as it is the most influential factor in injection molding.

Replication and Order

We must also decide on "replication" and "order". The replication indicates how many times process combinations will be tested (if at all). Replication can be partial or multiple; whereby, the partial is faster, but data is incomplete – not all process combinations are tested. Multiple means running the same combinations of factors and levels more than once to achieve higher confidence in the results (and to estimate error). "Order" refers to the order in which the selected conditions (combination of factors) are run; these should be randomized. It is often useful to reorder so that melt temp is only changed once. It is time consuming and difficult to know when true equilibrium is reached with regards to melt temperature changes. The thermocouple may indicate a change has been accomplished, but it is not located in the melt stream and temperature controllers may overshoot the setpoint before resuming normal control where heaters cycle on and off on a regular basis.

ANOVA

We have previously referred to the following concept (or derivation thereof):

$$\sigma^2{}_{total} = \sigma^2{}_{factors} + \sigma^2{}_{error}$$

In this case there are many separate variances for factors: each individual factor AND the interactions of factors.

In a typical ANOVA output table there various columns of data:

```
ANOVA for Factorial Model (Design Expert®; Stat-Ease, Inc)
Analysis of variance table [Partial sum of squares]
              Sum of              Mean          F
Source        Squares    DF       Square      Value    Prob>F
Model         6526.86     2       3263.43     39.83    0.0008
B             4482.86     1       4482.86     54.72    0.0007
C             2044.00     1       2044.00     24.95    0.0041
Residual       409.63     5         81.93
Cor Total     6936.50     7
```

The table is interpreted or can be calculated as follows ...
(÷, = & arrow symbols added to the standard data output)

```
              Sum of                Mean           F
Source        Squares      DF       Square       Value    Prob>F
Model         6526.86   ÷   2   =   3263.43   =  39.83    0.0008
B             4482.86   ÷   1   =   4482.86   =  54.72    0.0007
C             2044.00   ÷   1   =   2044.00   =  24.95    0.0041
                                  --------
Residual       409.63   ÷   5   =     81.93
Cor Total     6936.50       7
```

The F value is computed by dividing each Mean Square value by the residual (pure error ... aka within the subgroup variation). Most of the aforementioned is easy to follow except the Sum of Squares column; The SS_B & SS_T terms are as follows (the raw data from page 109 this DOE section, factor B "ANGLE" ... Y=response):

$$SS_B = \frac{\left(\sum B-\right)^2}{n} + \frac{\left(\sum B+\right)^2}{n} - \frac{\left(\sum Y\right)^2}{n}$$

$$SS_T = \sum\left(Y^2\right) - \frac{\left(\sum Y\right)^2}{n}$$

$$SS_B = \frac{\left(282.63\right)^2}{4} + \frac{\left(472\right)^2}{4} - \frac{\left(754.63\right)^2}{8}$$

$$SS_T = 78118.86 - \frac{754.63^2}{8}$$

$$SS_B = 19969.22 + 55696 - 71182.36$$

$$SS_T = 6936.50$$

$$SS_B = 4482.86$$

Your results might be slightly different if you try to re-create calcs above as the data on page 109 is rounded ... the SS_B & SS_T calcs above based on more digits (e.g. 58.3 was really 58.25 & 90.9 was really 90.875)

ANOVA (continued)

The "Residual" (error, noise, normal variation) of 409.63 comes from 6936.50 - 4482.86 - 2044; this is actually the calculated error term. As can be seen, the math can be difficult, but there are many DOE software programs to perform this math and display outputs as shown. Some DOE software programs may not give you a SS error term (as listed on previous page). This is because, all the degrees of freedom may be used to estimate various factors and interactions; when this occurs there will not be an error term (residual), and when there is no error term, there will not be an F ratio.

The F ratio is an indicator of statistical significance: higher value means greater statistical significance (the F ratio needs to be higher than the F distribution table value based on df for factor and residual error; use proper table for α risk). We can also look at the P value (Prob>F as listed previous page) to determine the percent chance that the same result could occur due to chance (we falsely reject the null hypothesis). The P value (Prob>F) is the result of an F probability distribution (FDIST function in Excel based on F Value, df in numerator & df in denominator). Calculates the probability that we are falsely rejecting the null hypothesis stating that the factor mean is different from error mean by amount indicated (factor is truly different than normal; thus, significant). e.g. factor C above: FDIST(24.95,1,5) = 0.0041 meaning there is a 0.41% chance that the F value 24.95 is caused by chance.

DOEs, DF and Error Terms

The total degrees of freedom (df) is one less than number of observations (n -1 ... it is n -1 because we use one df to estimate the grand average of the data set). Each time we calculate the significance of a factor or interaction we use one df; if we use them all, the residual or error will not be estimated, then we do not get the F-ratio as is missing from table below:

```
ANOVA for Factorial Model (Design Expert®; Stat-Ease, Inc)
Analysis of variance table [Partial sum of squares]
                Sum of              Mean        F
Source          Squares      DF     Square      Value    Prob>F
Model           6936.50      7      990.93
A               193.80       1      193.80
B               4482.86      1      4482.86
C               2044.00      1      2044.00
AB              1.22         1      1.22
AC              9.30         1      9.30
BC              141.33       1      141.33
ABC             63.99        1      63.99
Pure Error      0.000        0
Cor Total       6936.50
```

POOLING

There are various techniques that will yield enough data to generate this error term. The first technique is called pooling. Pooling involves deselecting from the analysis some factors or interactions. In order to exclude some factors we need to know they are not significant; we typically will look at a probability plot to determine points (factors) which follow a predictable pattern vs points well off the normal probability plot. Points "well off" are typically significant (such a graph is used on next page's analysis to select factors B & C only for study). The following ANOVA has factor A, and interactions AB, AC, BC, & ABC pooled (deselected) to estimate error (5 df); we get the following ANOVA (a probability plot generated by Design Expert® was used to select B & C - same as next page, but not shown here).

```
                Sum of              Mean        F
Source          Squares      DF     Square      Value    Prob>F
Model           6526.86      2      3263.43     39.83    0.0008
B               4482.86      1      4482.86     54.72    0.0007
C               2044.00      1      2044.00     24.95    0.0041
Residual        409.63       5      81.93
Cor Total       6936.50      7      7
```

DOEs, DF and Error Terms - MINITAB®

If we select all the terms for inclusion in the analysis model, we get:

```
MINITAB® Analysis of Variance for Yavg (coded units)
Source              DF    Seq SS    Adj SS    Adj MS   F   P
Main Effects         3   6720.66   6720.66   2240.22   *   *
  A                  1    193.80    193.80    193.80   *   *
  B                  1   4482.86   4482.86   4482.86   *   *
  C                  1   2044.00   2044.00   2044.00   *   *
2-Way Interactions   3    151.85    151.85     50.62   *   *
  A*B                1      1.22      1.22      1.22   *   *
  A*C                1      9.30      9.30      9.30   *   *
  B*C                1    141.33    141.33    141.33   *   *
3-Way Interactions   1     63.99     63.99     63.99   *   *
  A*B*C              1     63.99     63.99     63.99   *   *
Residual Error       0       *         *         *
Total                7   6936.50
```

Note: Just as was the case with the Design Expert® software from Stat-Ease, Inc, (previous page) we did not get an estimate of error or noise; thus, no F or P values to indicate significance. We can look at the Probablity plot below and then select terms to be included vs excluded in the analysis to be done ... only terms B & C below are shown to be significant.

The Normal plot below made using Minitab v16.1 ... Create by selecting Stat, DOE, Factorial, Analyze ... Select Response (Yavg in this example) ... Select Graphs from the Analyze options: turn on Normal Plot & Effects Plots, etc.

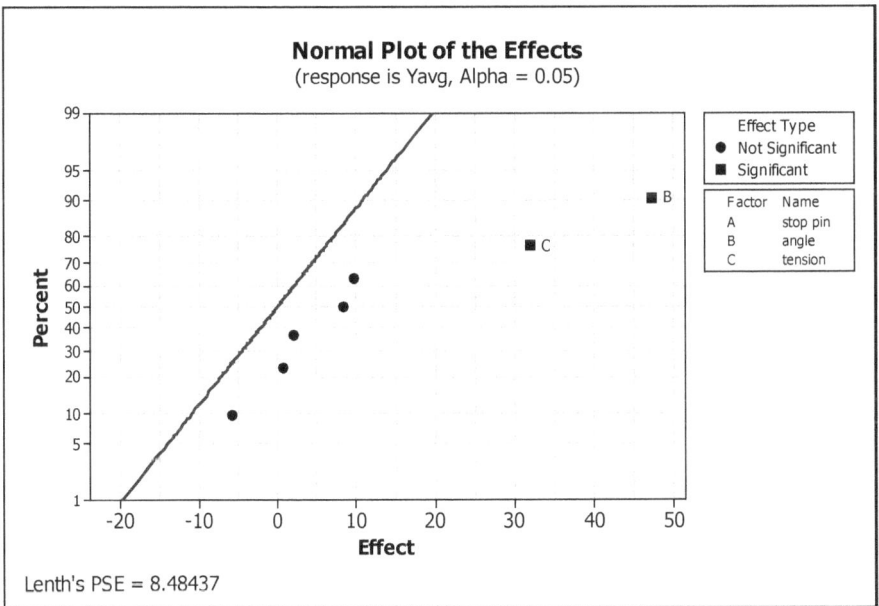

Factors B & C are exceeding Lenths ME (margin of error) ... these are typically picked or identified for selection based on graphical deviation from the bulk of data.

DOEs, DF and Error Terms - MINITAB® (continued)

Analyze in Minitab®, select menu options as follows:
Stat, DOE, Factorial, Analyze Factorial Design ... select Response to Analyze (Yavg in this example) ... select terms as exemplified in screen below (identifying which terms are not selected for analysis allows them to be used for estimating error).

The selected terms were chosen based on the Normal Plot classifying them as Significant (previous page).

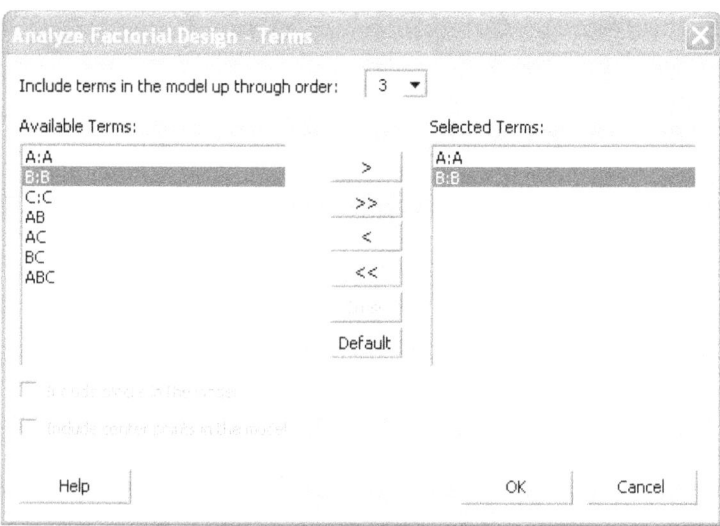

Source	DF	Seq SS	Adj SS	Adj MS	F	P
Main Effects	2	6526.9	6526.9	3263.43	39.83	0.001
B	1	4482.9	4482.9	4482.86	54.72	0.001
C	1	2044.0	2044.0	2044.00	24.95	0.004
Residual Error	5	409.6	409.6	81.93		
Total	7	6936.5				

DOEs, DF and Error Terms - MINITAB® (continued)

NOTE: this is same data from statapult experiment on earlier pages ... we got same result whereby factor B is the most significant followed by factor C. This data also looks the same as was derived from Design Expert® software; thus, there are various programs to analyze DOE data.

BLOCKING, REPLICATION & CENTER POINTS ADD DF
Other techniques used to estimate error involve designing the experiment such that an abundance of df is created; more observations equal more df (degrees of freedom). This can be accomplished by replicating the experiment or having multiple center points. Replicated factor combinations or replicated center points would yield same response value if variation was non-existent; thus, differences yield error (normal variation).

Blocking involves replicating some factors within a blocked factor such as two different machines or same machine and two different resins or same experiment at two different times. When blocking in MINITAB®, take care to select two blocks AND two replicates; two blocks with one replicate results in blocks containing partial replicates.

How many degrees of freedom are enough? Approximately five or six df should be sufficient to predict the error term. Do not pool if there are already six df.

Fully replicated experiments (meaning two replicates or two runs of same factor combinations) result in an ample amount of df. Center points will add some as well, but may use one to check for curvature in the data. It is often useful to arrange center points whereby, they are located at start, middle and end of run. Center points can be used to check for curvature and/or used to increase the df available for estimating error. There can be many center points.

DOE Planning Template

All or most of the following should be done as a team:
- <u>List the objectives</u>; background info & assumptions
- List the chosen responses, factors and levels
- Assign a constant value for factors not chosen
- Decide on number of observations (shots) to be collected; will they be used for mean or standard deviation as a response?
- Decide on number of replicates
- Determine if any factors should be blocked
- Decide if and how many center points will be included; location such as start, middle, end, etc (what is purpose for center points)
- Decide if experiment will yield sufficient df to estimate error ... adding center-points and running these center points 2-3 times provides a good estimate of error ... estimates of error are needed to help compute statistical significance
- Is measurement system capable (acceptable Gage R&R values)

See next page for a completed sample DOE Planning Template.

DOE Planning - Sample Template

DOE Planning Template	
Date:	August 15, 2002
Name:	Mold Optimization DOE
Title:	C7 Flatness Optimization
Objective:	Identify cycle time effects on certain CTFs including visual appearance. Identify real time SPC monitoring outputs. Identify best control dimension for SQC checks and best SPC opportunities.
Background:	Desired cycle time is 20 seconds maximum for best productivity. Resin known to have significant variation. Packing known to effect flatness.
Assumptions:	Resin lot testing will be typical of future resin lots

Response Variables:	Measurement Techniques:
Length: 44.250 ± 0.150	Gage xx110B
Width: 25.578 ± 0.125	Gage xx112B
Height: 37.750 ± 0.150	RJG eDart & Kistler piezoelectric transducers
Flatness: locations 1 & 2	
Data log cav psi + other	Machine printouts & RJG instrumentation

Factors:	LO	CENTER	HI
Resin - 2 blocks	536672...5572 poise	551644...6410 poise	
Dryer temp	Constant 260° F at hopper inlet		
Dryer time	Constant 6+ hrs		
Resin temp (front zone barrel/ascending profile)	Constant 545° F (ascending profile setup D2211)		
HR System Manifold	Same as barrel front zone		
HR system probes	565-595°F (setup D2211 in HR controller for balance)		
Mold A temp	Constant 80° F		
Mold B temp	60°F	70°F	80° F
Mold C temp (slides)	Constant 60° F		
Inj fill time	Constant 1.3 sec (93.56% transfer weight)		
nozzle diameter	0.375 - same as mold		
cooling time (as it affects cycle time)	7	9	11
pack time	Constant 3.5 sec - profile X - setup D2211		
pack psi (cav psi target peak)	9720	11220	12720
shot size	Constant 0.90 inches		
screw RPM	Constant 165 rpm		
back psi	700 ppsi (plastic)		
decomp	Constant 0.200		

Noise:	Control Method:
Resin Variation	Blocked
Measurement error	MSE performed...use same metrologist
Machine variation	Data Log key parameters to ID changes from norm

Replication	Two replicates (one per block)
Centerpoints	One per block
Blocks:	Two blocks
Randomization	Fully randomized
Design Matrix	2^3 full factorial replicated
Analysis:	Mini-Tab
Est time/cost:	

The gray shaded cells will be the factors and levels tested in this DOE

ANOVA w/ Blocking, Replicates & C.P.s

The ANOVA table on next page (from MINITAB®) is for the data set seen much earlier in section on Effective SQC (DOE Data Example # 1 - p.47), and the DOE planning template on previous page. There is a block on resin lot. Ample df (8 left over) yield an estimate of residual error.

The main effects are highly significant; specifically factor A (peak psi) and factor C (cycle time). Both are indirect relationships meaning: higher psi and longer time yield flatter parts (lower flatness values). Cycle time is a stronger effect for the range tested vs peak psi for range tested. The data indicates the factors to be much more significant than the block (resin lot) ... 307.46 vs 6.92 (threshold values for the 6.92 block can be viewed by comparing F distribution table value at 5% for df values of 1 in numerator and 8 in denominator = 5.32) ... we are just barely beyond the threshold value, but quite small compared to the 307.46.

The real value of this DOE ANOVA output (shown on next page) is the table of coefficients for C7 Flatness (shaded in gray at bottom of next page).

We can also create Main Effects Plots as a graphical indication of how to adjust process to get best flatness. Create effects plots by selecting: Stat, DOE, Factorial, Factorial Plots, Main Effect, Setup, ... select Response ... select Factors, Options, OK. There are also cube plots, etc which can serve as a roadmap for reference when trying to process correct certain dimensions.

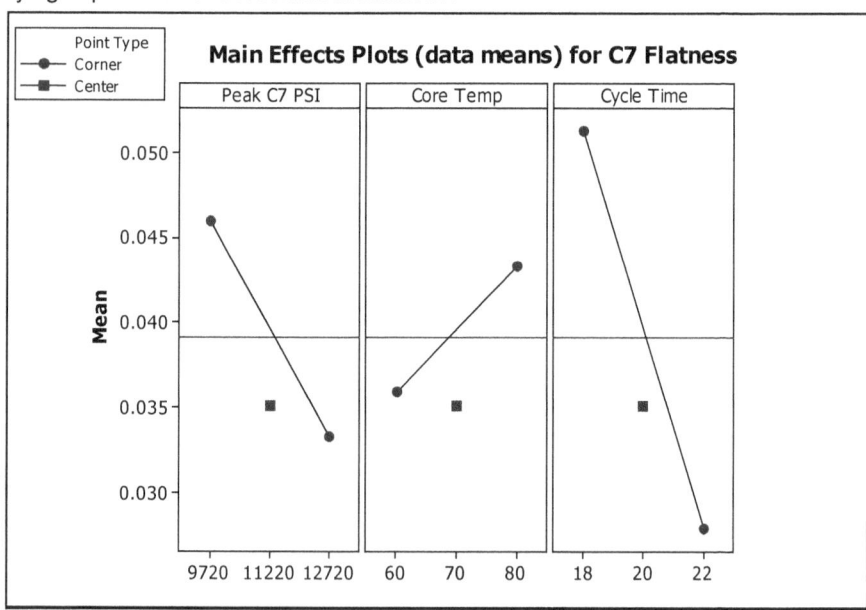

These plots are very helpful to process engineers, but the coefficients mentioned above which will be used in a regression equation are even more useful ... see next two pages.

```
Estimated Effects and Coefficients for C7 Flatness (coded units)

Term                            Effect      Coef   SE Coef       T       P
Constant                                 0.03959  0.000455   87.08   0.000
Block                                   -0.00113  0.000429   -2.63   0.030
Peak C7 PSI                    -0.01271 -0.00636  0.000455  -13.98   0.000
Core Temp                       0.00744  0.00372  0.000455    8.18   0.000
Cycle Time                     -0.02336 -0.01168  0.000455  -25.69   0.000
Peak C7 PSI*Core Temp           0.00089  0.00044  0.000455    0.98   0.358
Peak C7 PSI*Cycle Time          0.00379  0.00189  0.000455    4.17   0.003
Core Temp*Cycle Time           -0.00486 -0.00243  0.000455   -5.35   0.001
Peak C7 PSI*Core Temp*Cycle Time -0.00156 -0.00078  0.000455  -1.72   0.124
Ct Pt                                   -0.00449  0.001364   -3.29   0.011
R-Sq = 99.20%     R-Sq(pred) = 94.42%    R-Sq(adj) = 98.30%

Analysis of Variance for C7 Flatness (coded units)
Source                            DF      Seq SS      Adj SS      Adj MS
Blocks                             1  0.00002289  0.00002289  0.00002289
Main Effects                       3  0.00305092  0.00305092  0.00101697
  Peak C7 PSI                      1  0.00064643  0.00064643  0.00064643
  Core Temp                        1  0.00022127  0.00022127  0.00022127
  Cycle Time                       1  0.00218323  0.00218323  0.00218323
2-Way Interactions                 3  0.00015511  0.00015511  0.00005170
  Peak C7 PSI*Core Temp            1  0.00000315  0.00000315  0.00000315
  Peak C7 PSI*Cycle Time           1  0.00005738  0.00005738  0.00005738
  Core Temp*Cycle Time             1  0.00009458  0.00009458  0.00009458
3-Way Interactions                 1  0.00000977  0.00000977  0.00000977
  Peak C7 PSI*Core Temp*Cycle Time 1  0.00000977  0.00000977  0.00000977
Curvature                          1  0.00003590  0.00003590  0.00003590
Residual Error                     8  0.00002646  0.00002646  0.00000331
Total                             17  0.00330105

Source                               F       P
Blocks                            6.92   0.030
Main Effects                    307.46   0.000
  Peak C7 PSI                   195.44   0.000
  Core Temp                      66.90   0.000
  Cycle Time                    660.06   0.000
2-Way Interactions               15.63   0.001
  Peak C7 PSI*Core Temp           0.95   0.358
  Peak C7 PSI*Cycle Time         17.35   0.003
  Core Temp*Cycle Time           28.59   0.001
3-Way Interactions                2.95   0.124
  Peak C7 PSI*Core Temp*Cycle Time 2.95  0.124
Curvature                        10.85   0.011

Estimated Coefficients for C7 Flatness using data in uncoded units
Term                                Coef
Constant                         0.581682
Block                           -0.00112778
Peak C7 PSI                     -5.53917E-05
Core Temp                       -0.00337255
Cycle Time                      -0.0248670
Peak C7 PSI*Core Temp            5.50417E-07
Peak C7 PSI*Cycle Time           2.45417E-06
Core Temp*Cycle Time             0.000170625
Peak C7 PSI*Core Temp*Cycle Time -2.60417E-08
Ct Pt                           -0.00449375
```

Regression Coefficients for Prediction

From the ANOVA on previous page, the upper 2/3s of ANOVA output is for coded units meaning factors are treated as -1 to +1; this provides a normalized scale so that coefficients in "Coef" column can be easily scanned to see which factor has greatest relative effect.

Factors Cycle Time and Peak C7 PSI have a very strong effect, but all three main factors are significant. If we want to predict the process that would result in a desired 0.025 flatness, we would use coefficients from bottom 1/3 of table -- uncoded units ... highlighted in gray on previous page.

At bottom of this page, these coefficients are used to calculate the process needed using a separate spreadsheet program. The required cycle of 19 is set then, Peak PSI and Core Temp varied to see combo that produces 0.025 or less. Once this is entered into EXCEL, we can do what if analysis or use it's "Solver" (w/ limits) to identify best process for lowest flatness.

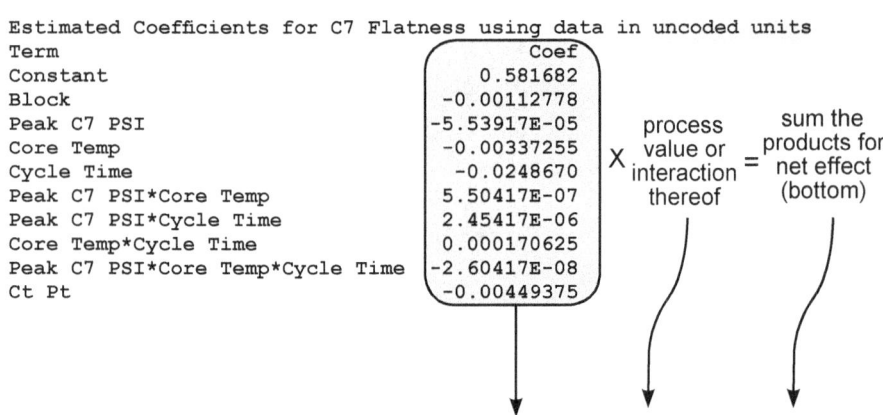

```
Estimated Coefficients for C7 Flatness using data in uncoded units
Term                                     Coef
Constant                             0.581682
Block                               -0.00112778
Peak C7 PSI                         -5.53917E-05
Core Temp                           -0.00337255
Cycle Time                          -0.0248670
Peak C7 PSI*Core Temp                5.50417E-07
Peak C7 PSI*Cycle Time               2.45417E-06
Core Temp*Cycle Time                 0.000170625
Peak C7 PSI*Core Temp*Cycle Time    -2.60417E-08
Ct Pt                               -0.00449375
```

process value or interaction thereof X = sum the products for net effect (bottom)

	A	B	C	E	F
1		Term	Coef	Process	
2		Constant	0.581682	1	0.581682
3		Block	-0.00112778		0.000000
4	Peak C7 PSI	A	-5.54E-05	14085	-0.780192
5	Core Temp	B	-0.00337255	60	-0.202353
6	Cycle Time	C	-0.024867	19	-0.472473
7		A*B	5.50E-07	845100	0.465157
8		A*C	2.45E-06	267615	0.656772
9		B*C	0.000170625	1140	0.194513
10		A*B*C	-2.60E-08	16056900	-0.418149
11		Ct Pt	-0.00449375		0.000000
12	C7 Flatness	>>>>>>> calculated >>>>>>>>>>>>			0.02496

Partial Replication & Aliasing

While multiple replicates can yield additional df for a better error term, partial replication does not and results in confounding. Confounding results when some terms or interactions are aliased or confounded with other terms or interactions. When an alias occurs, the effect of one factor or interaction cannot be mathematically separated from another; this will occur when there is a partial or fractional factorial experiment.

Why would we ever want to perform a partial replication (fractional factorial)? The answer is to save time and money on a smaller experiment, but we must understand the aliasing present.

There is a principle known as sparsity of effects which means that higher order interactions are unlikely to be significant; thus, we likely don't mind if three and four level interactions are aliased with some factors or two factor interactions. Two factor interactions that are aliased with other two factor interactions can sometimes be addressed by viewing the alias patterns (supplied by most software). If there is an expected or predicted possible interaction that is aliased with another, then sometimes the terms can be reassigned so that these two level interactions are now aliased with something else (e.g. make pack pressure factor C instead of B, etc...requires DOE planning). Pack may interact with fill time for a synergistic effect, but likely not cycle time; as you assign terms with partially replicated experiments, you must predict the possibility of an interaction or if aliased factors are significant do follow-up experiments to verify.

Consider the following 2^4 experiment but done as a ½ fraction; this results in a resolution IV experiment with eight runs instead of 16. In resolution IV: single factors are aliased with three factors and two factors aliased with other two factors: $1+3=4$ and $2+2=4$; thus resolution IV. Note: we want to <u>avoid resolution III experiments where single factors are aliased with two factors</u>.

As we look at the columns, <u>some columns have exact same pattern as other columns</u>; thus, cannot be differentiated from each other.

SO	A	B	C	D	AB	AC	AD	BC	BD	CD	ABC	ABD	ACD	BCD	ABCD
1	-1	-1	-1	-1	1	1	1	1	1	1	-1	-1	-1	-1	1
2	1	-1	-1	1	-1	-1	1	1	-1	-1	1	-1	-1	1	1
3	-1	1	-1	1	-1	1	-1	-1	1	-1	1	-1	1	-1	1
4	1	1	-1	-1	1	-1	-1	-1	-1	1	-1	-1	1	1	1
5	-1	-1	1	1	1	-1	-1	-1	-1	1	1	1	-1	-1	1
6	1	-1	1	-1	-1	1	-1	-1	1	-1	-1	1	-1	1	1
7	-1	1	1	-1	-1	-1	1	1	-1	-1	-1	1	1	-1	1
8	1	1	1	1	1	1	1	1	1	1	1	1	1	1	1

Most software will supply the matches (alias structure) as follows:
A + BCD (meaning A = BCD) ... other confounding includes:
B + ACD ... C + ABD ... D + ABC ... AB + CD ... AC + BD ... AD + BC

Each interaction is just the multiplication of the 1s or -1s (e.g. -1 x -1 = 1, etc).

Type I vs Type II Error

A Type I error is made when we call something significant when it is not (reject a null hypothesis that is true). This is a mistake that errs on the conservative side in that it protects the customer, but adds cost to the molder or supplier.

The risk for committing a Type I error is called an alpha (α) risk. We want the P value to be less than the set α value or maximum acceptable α risk. This is often set at 0.05 (5%). If the product is flower pots, the acceptable α risk might be set higher at 0.10, and if we are making molded products used in critical care medical applications, the maximum risk might be set at 0.01 (1%). We would look at P value relative to this α risk when deciding when something is statistically significant or not.

A Type II error is made when we say something is not significant when it really is. This mistake is more detrimental to the customer. The risk for committing a Type II error is called the β risk.

Ideally we want to avoid making either type error as they both effect cost (either directly or as consequence to lost customer satisfaction).

Calculating Sample Size to Avoid Type I & II Errors
A spreadsheet can be constructed as shown below. In this instance, we want to detect a difference of 0.045 (33% of distance to nearest specification limit ... 37.900 minus 37.765 multiplied by 0.33). A separate data set was examined with constant process conditions whereby the σ was 0.0184 (but this might have been the identified variation found in the measurement system). We set the α & β risks at 0.05 (5%). In this case we only need to measure four shots to compute our average for H-avg.

If the mean had been 37.815 (closer to the USL); the calculated sample size would have been 9 because we would need to detect a smaller delta of 0.028 (one third the distance to nearest spec limit ... 0.028 instead of 0.045).

	A	B	C	D	E	F	G	H
1		Sample size	INPUTS					
2		1 sided test	V V V V					Suggested
3	δ	delta	0.045	What size difference do you want to detect? Suggest set at no higher than 1/3 min distance to closest spec limit >>>				0.045
4	α	alpha risk	0.05	What alpha risk are you willing to take? Type I error - % chance that we call something significant when it really is not (typ 5-10%).				0.05
5	β	beta risk	0.05	What beta risk are you willing to take? Type II error, Fail to say it is significant when it is; could result in failures outside the facility (typ 5-20%).				0.05
6	σ	sigma	0.0184	How much variation do you expect in the data?......What has been the typical std dev? Suggest max std dev be used >>>				0.0184
7	N	Calculated sample size	4	Calculated sample size based on formula below and misc shaded inputs (columns A & C): $\delta, \alpha, \beta, \sigma$				NA
8			H-avg					
9		Z values	7.337					
10		mean	37.765					
11		sigma	0.0184					
12		USL	37.900					
13		LSL	37.600					
14		min dist to spec	0.135					
15	33%		0.045					

$$N = 2(Z_\alpha + Z_\beta)^2 \times \frac{\sigma^2}{\delta^2}$$

C7=ROUND(2*((NORMSINV(C4)*-1)+(NORMSINV(C5)*-1))^2*(C6^2/C3^2),0)

EVOP

After the process development and qualification, there are still opportunities for continuous improvement. These activities may be run as EVOP experiments: meaning an evolutionary operation. An EVOP experiment is an experiment performed simultaneous to ongoing normal production; thus, all product produced will be saved and sold, but data will be collected and analyzed. In order to not adversely effect part requirements, only very small changes will be applied. An EVOP experiment, is much like a 2^2 DOE experiment with a centerpoint, but replicated several times with the average of replications used for analysis. The signal to noise ratio is small; thus, the reason for multiple runs or replications. During the qualification of parts, a nominal center process is established with upper and lower limits assigned based on window study data accumulated during the process development. During the EVOP experimentation, a small window inside the normal process window will be explored, but with many parts run at each condition. Typically, there is some DOE data or profound process knowledge that identifies which factors are important. During the EVOP phase, the <u>normal inspection and specification limits would be applicable with compliance verified</u>. Control limits may be violated as there is controlled variation being applied. ANOVA or regression analysis can check for statistical significance, but often a square plot as shown below can be used to see if the means have moved as desired. Checking and comparing the average variance at each condition is suggested to guard against a desired mean shift accompanied by increased variation. Frequently there will be some signal present which favors one corner to migrate toward. Consider the following whereby a target dimension of 1.670 was 0.003 low ...over several months, the dimension could be adjusted up to the desired 1.670. Cycle had minimal effect; thus, mold temperature was added to the pack pressure for investigation.

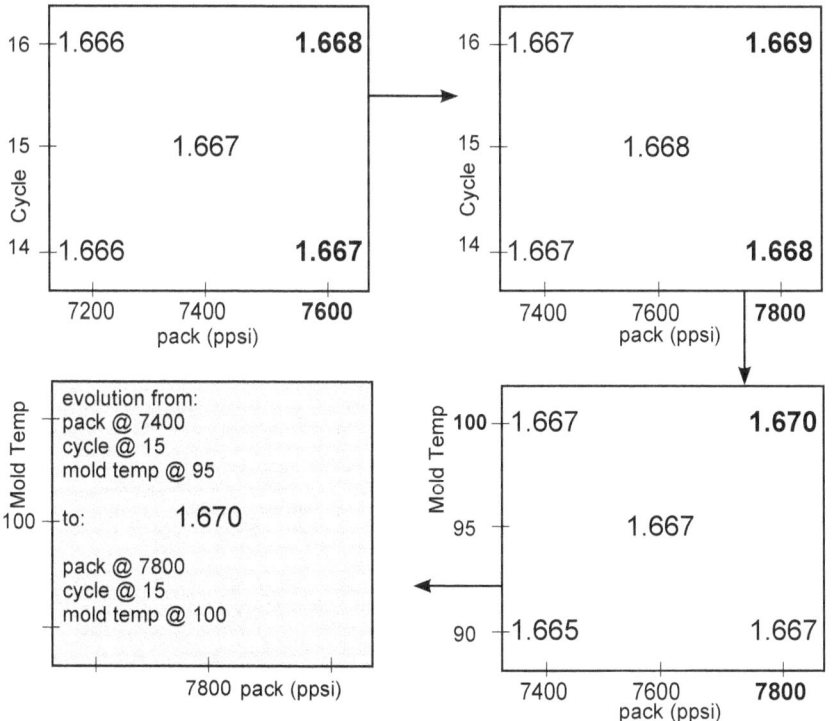

Process Validation

The process validation allows the molder to obtain and interpret data which will determine the process to be capable of meeting requirements – internal and external.

There are typically three phases to this validation:
1. IQ – installation qualification; includes checks for basic capability and functionality of molds, machines and equipment.
2. OQ – operational qualification; includes checks to determine process nominalization (centering); establish statistically based control limits. With specification limits, process mean and variation established, an initial risk assessment can be made.
3. PQ – performance qualification; includes longer term runs to check for process stability and capability.

Often times there are window studies (small DOEs) performed in the OQ phase. These window studies start as functional windows whereby limits for flash and shorts are identified; then a process window is identified (all process combinations within this window must yield parts within specification limits; ideally the middle 50% of specification range). A center process will be established.

During the PQ phase the same center process will be used: this should be checked against the center process results achieved during the OQ phase when process was finalized. A one way ANOVA can be used to verify if the two runs yield the same performance.

Typically in a DOE, we are using the ANOVA to identify and quantify the means which are different as caused by different process factors. We will now perform the one way ANOVA to verify the means are _not_ statistically different; thus, indicating that the process is robust and repeatable at two different times. In Minitab® select: Stat, ANOVA, One-way ... select Response ... select Factor ... select Graphs - Individuals & Boxplots & residuals. The F ratio @ 0.97 below indicates means to be no different statistically; with df @ 1 & 148 the F ratio would need to be greater than 3.84 to be considered significant w/ α @ 0.05).

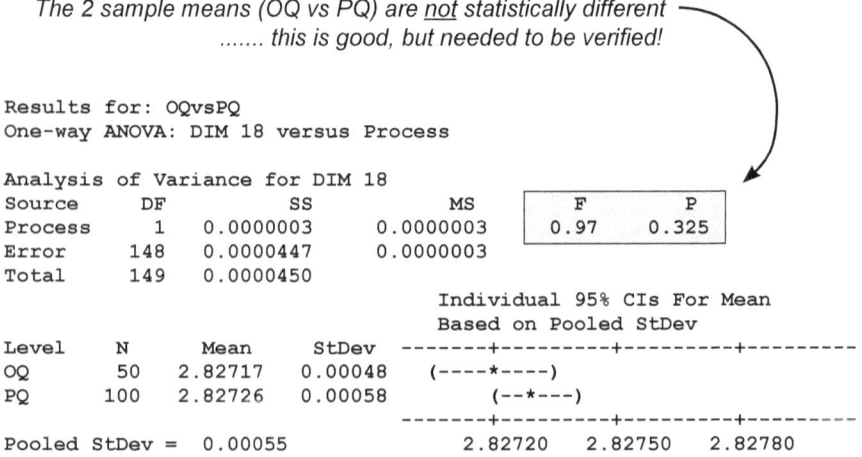

The 2 sample means (OQ vs PQ) are _not_ statistically different
.... this is good, but needed to be verified!

```
Results for: OQvsPQ
One-way ANOVA: DIM 18 versus Process

Analysis of Variance for DIM 18
Source      DF        SS          MS          F        P
Process      1    0.0000003   0.0000003    0.97     0.325
Error      148    0.0000447   0.0000003
Total      149    0.0000450

                                  Individual 95% CIs For Mean
                                  Based on Pooled StDev
Level    N      Mean     StDev    -------+---------+---------+---------+---------
OQ      50    2.82717   0.00048   (----*----)
PQ     100    2.82726   0.00058          (--*---)
                                  -------+---------+---------+---------+---------
Pooled StDev =  0.00055              2.82720   2.82750   2.82780
```

Process Validation (OQ vs PQ)

The one way ANOVA from previous page is re-listed below so as to display with the boxplots.

```
Results for: OQvsPQ
One-way ANOVA: DIM 18 versus Process

Analysis of Variance for DIM 18
Source      DF         SS           MS        F        P
Process      1     0.0000003    0.0000003    0.97    0.325
Error      148     0.0000447    0.0000003
Total      149     0.0000450
                                    Individual 95% CIs For Mean
                                    Based on Pooled StDev
Level     N      Mean      StDev   -------+---------+---------+---------
OQ        50   2.82717    0.00048  (----*----)
PQ       100   2.82726    0.00058        (--*---)
                                   -------+---------+---------+---------
Pooled StDev =   0.00055           2.82720   2.82750   2.82780
```

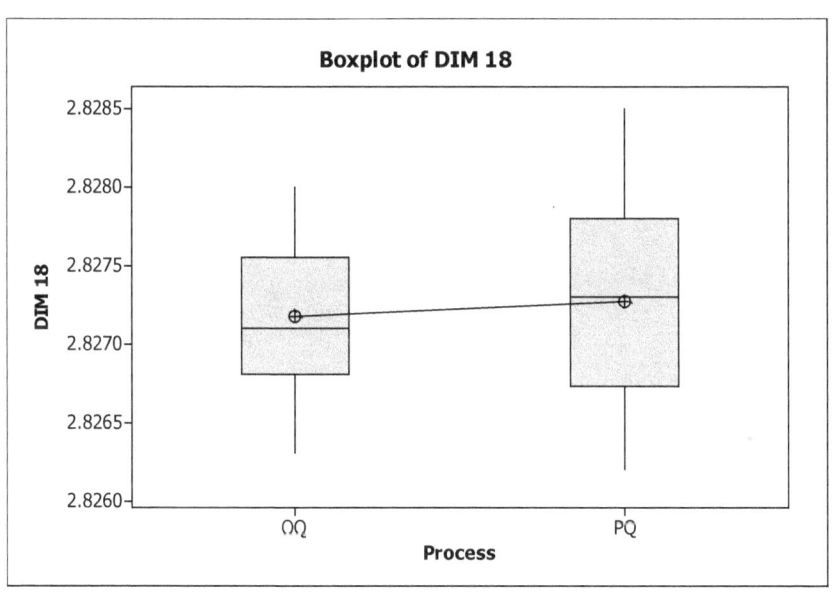

Boxplot of DIM 18

Process Validation (PQ Boxplot)

On the previous page, we verified that the current process capability run was no different (i.e. statistically via one way ANOVA) than the center process run performed during earlier process development.

The data below will review the latest process validation run (50 shots of two cavities collected over eight hours). This mold is a two cavity mold with all data grouped together; thus, cavity to cavity variation is present and can be seen in the boxplot below. The duration of this type validation run does vary with different molders. When deciding the duration, consider the potential sources of variation that needs to be tested such as ambient temperature, dryer equilibrium affected by residence time, shift personnel, etc.

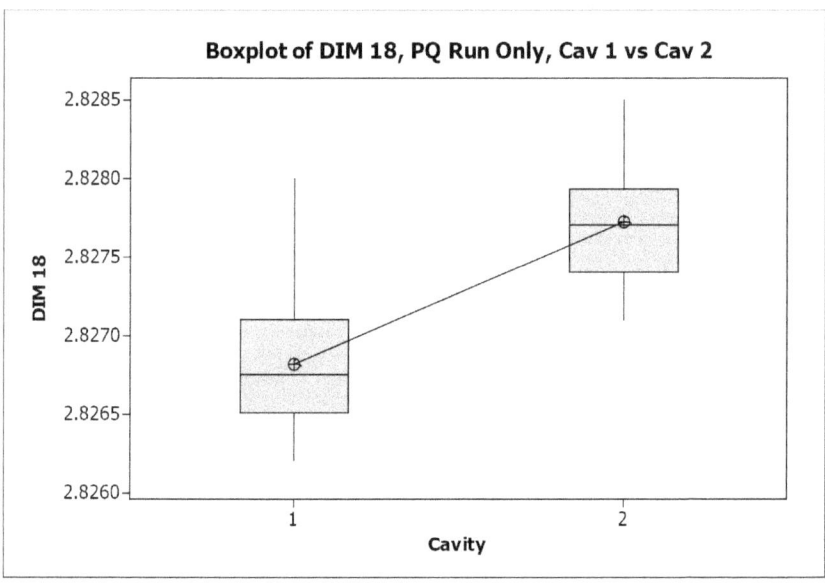

The boxplot above made in MINITAB® as follows; select: Stat, Anova, One way ... select Response (Dim 18) ... select Factor (Cavity) ... select Graphs ... select Individual value plot & Boxplots, and can select Residual Plots as desired.

Process Validation (PQ Capability)

After reviewing the expected ppm defective, we can see the process performance is very good, but not great as the Z score is 4.71 (Ppk at 1.57). An expected 1.23 ppm defective is <u>predicted</u>. It should be noted that this is only 0.00012%; thus, very good. If this product is for critical care medical product: additional centering might be suggested. Often times the cost to correct is compared to cost and impact of non-compliance failures. This process may be able to be adjusted via regression equations without expensive tool rework, but a process correction will retain the cavity differences, and may change more than the needed features (maybe adversely affecting some).

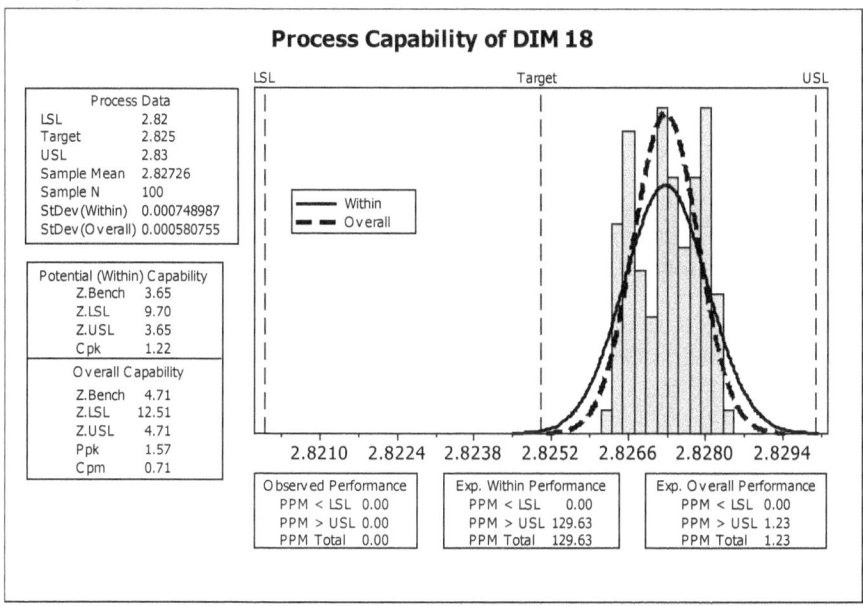

We can make a process capability graphic/analysis like shown above in MINITAB® as follows; Select: Stat, Quality Tools, Capability Analysis, Normal ... select dimension for study, set Subgroup Size at 2, enter spec limits ... select options and enter target, select entries for perform analysis (check both), display PPM, display Benchmark Z's (or select capability stats) ... select Estimate and select Pooled standard deviation.

Using Metrics

There are advantages to having metrics related to our processes:
1. Identify change in performance (i.e. improvement or decline).
2. Identify need for change (followed by improvement plan & action).
3. Establish priorities (derive best return for effort).
4. Sense of accomplishment is motivating (see below).

Metrics can be:
5. Time: delivery, response, completion vs. schedule, etc.
6. Financial: cost, cost reduction, relative to budget, ROI, etc.
7. Performance: efficiency, yield, utilization, safety, rework, etc.
8. Product performance related: dimensional, physical, etc.

In order for metrics to be tracked via a scorecard, they need to be characterized by the following (see scorecard on next pages):
9. Numerical (even attribute data such as yes/no or good/bad can be counted to become number data, but a larger sample size will be needed to be useful). Compare a completion date that is or is not on time vs. actual days to complete.
10. Have numerical specification limits listing requirements: upper and lower limits.
11. Have accumulated data so that a standard deviation can be computed.

The actual measurement does not have much value until we use it for feedback. This can be motivating in that a sense of accomplishment is derived from improvements made ... or can be motivating to know what improvements are needed. Due to the competitive nature that is present in most people, they like to know how they measure up.
12. The measurement scale needs to have sufficient resolution or discrimination capability so that performance and/or changes can be accurately depicted.
13. The measurement feedback needs to be timely relative to the needs for change: this could be hourly, daily, monthly, etc depending on the need for intervention to make corrections.
14. The measurements should be sufficiently detailed so that improvements can be made on an individual level. If an organization is measured on delivery and the performance is mid level, there are likely some individuals who are high level that may be carrying low level performers. A struggling custom molder may have highly profitable jobs that are carrying the low margin jobs ... this often results in price increases across the board which can result in losing the high margin business and keeping the low margin business if individual job performance is not known.

Managers at all levels should accept responsibility to manage. Frequently, managers rely on the workers to produce and measure themselves whether in accounting, engineering, etc (self measurement is good, but needs to be reviewed by others to supply feedback). Individuals can become conflicted knowing that the performance needs to be improved, but they have to be the ones to expend extra effort, AND they don't know if the management sees, cares or appreciates such effort. Managers get tasks accomplished thru people; the accomplishments can be substantial or mediocre depending on their approach to motivation: using scorecards, training, mentoring, rewards, etc.

Statistically Based Scorecard

The data table below and radar graph (next page) identify the Z score relative to a target of 6. The data table will provide specifics, and permit what if calculations: what if LSL or USL changed, what if variation is reduced or what if centering improves.

The scorecard can also be used in other ways besides dimensional compliance; such as cost, delivery, quality, etc ... requires number values for targets, spec limits, & std dev. When we get away from dimensions and more toward cost, delivery, scrap, etc, we may choose to rename the term centering to desirability or degree of optimization.

	G	H	I	J	K	L	M	N	O
1	Part	Weight	centering	desirability (d)	Best	Risk (TDU)	Part	Risk Z	Best
2	Valve	2	0.83	0.68	1	0.000934	Valve	3.11	6
3	Base	1	0.71	0.71	1	0.000000	Base	6.00	6
4	Spinner	4	0.54	0.08	1	0.000000	Spinner	5.05	6
5	Composite wtd d = >>			0.63					

Cell I2 =N37 ... L2=Q37 ... J2 =IF(ISBLANK(H2), "", I2^H2)
Cell N2 =IF(L2<0.000000001, O2, ROUND(NORMSINV(L2), 2)*-1)
Cell J5 =PRODUCT(J2:J4)^(1/(SUM(H2:H4)))

	F	G	H	I	J	K	L	M	N	O	P	Q	R
32	Valve	Units	Target	LSL	USL	Weight	Mean	Sigma	centering	ZLSL	ZUSL	DPU	PPM
33	1	inch	4.250	4.245	4.255	1	4.2493	0.00136	0.85	3.125	4.23	0.000901	900.8
34	2	inch	4.250	4.245	4.255	1	4.2511	0.00093	0.78	6.595	4.22	0.000012	12.4
35	3	inch	4.250	4.245	4.255	1	4.2488	0.00093	0.76	4.108	6.70	0.000020	19.9
36	4	inch	4.250	4.245	4.255	1	4.2496	0.00093	0.92	4.973	5.84	0.000000	0.3
37							TOTAL (assumes equal wtg) >>>		0.83		SUM >>>	0.000934	933.5
39	Base	Units	Target	LSL	USL	Weight	Mean	Sigma	centering	ZLSL	ZUSL	DPU	PPM
40	1	inch	8.500	8.450	8.550	1	8.5121	0.00234	0.76	26.538	16.20	0.000000	0.0
41	2	inch	8.500	8.450	8.550	1	8.5153	0.00175	0.69	37.314	19.83	0.000000	0.0
42	3	inch	8.500	8.450	8.550	1	8.5112	0.00188	0.78	32.553	20.64	0.000000	0.0
43	4	inch	8.500	8.450	8.550	1	8.5191	0.00110	0.62	62.818	28.09	0.000000	0.0
44							TOTAL (assumes equal wtg) >>>		0.71		SUM >>>	0.000000	0.0
46	Spinner	Units	Target	LSL	USL	Weight	Mean	Sigma	centering	ZLSL	ZUSL	DPU	PPM
47	1	inch	0.250	0.247	0.253	1	0.2512	0.00034	0.60	12.310	5.23	0.000000	0.1
48	2	inch	0.250	0.247	0.253	1	0.2515	0.00029	0.50	15.734	5.24	0.000000	0.1
49	3	inch	0.250	0.247	0.253	1	0.2516	0.00027	0.47	17.358	5.28	0.000000	0.1
50	4	inch	0.250	0.247	0.253	1	0.2512	0.00024	0.60	17.355	7.44	0.000000	0.0
51							TOTAL (assumes equal wtg) >>>		0.54		SUM >>>	0.000000	0.2

Cell N33 =IF(ISBLANK(H33),"", IF((J33-L33)<(L33-I33), 1-(L33-H33)/(J33-H33), 1-(H33-L33)/(H33-I33)))
Cell N37 =PRODUCT(N33:N36)^(1/(SUM(K33:K36)))
Cell O33 =IF(ISBLANK(I33),"", IF(M33<>0,(L33-I33)/M33,""))
Cell P33 =IF(ISBLANK(J33),"", IF(M33<>0,(J33-L33)/M33,""))
Cell R33=IF(NOT(ISBLANK(H33)), 1000000*(IF(ISBLANK(J33),0, IF(M33<>0,1-NORMSDIST(O33),0)) ꟾ IF(ISBLANK(K33),0, IF(M33<>0,1-NORMSDIST(P33),0))),"")

Cell J5 is a weighted "d" whereby cells J2,3 & 4 are weighted - shown for instruction purposes only as this "d" is for tool nominalization and weights would typically be equal.

Tables above use the convention of black for labels or fixed values, blue for user inputs and red for formulas calculating data.

Composite Desirability and Radar Graph

The weight (column H) makes no difference if all the same (different weights applied in this example for purposes of instruction). If the categories were performance values for quality, cost, delivery, scrap, etc for different molds or programs then the weighting would likely be varied. The total composite weight is based on following formula (d=desirability or degree of optimization):

$$d_+ = \left(d_1^{w1} \times d_2^{w2} \times d_3^{w3} \right)^{(1/(w1 + w2 + w3))}$$

	G	H	I	J	K	L	M	N	O
1	Part	Weight	centering	desirability (d)	Best	Risk (TDU)	Part	Risk Z	Best
2	Valve	2	0.83	0.68	1	0.000934	Valve	3.11	6
3	Base	1	0.71	0.71	1	0.000000	Base	6.00	6
4	Spinner	4	0.54	0.08	1	0.000000	Spinner	5.05	6
5	Composite wtd d = >>			0.63					

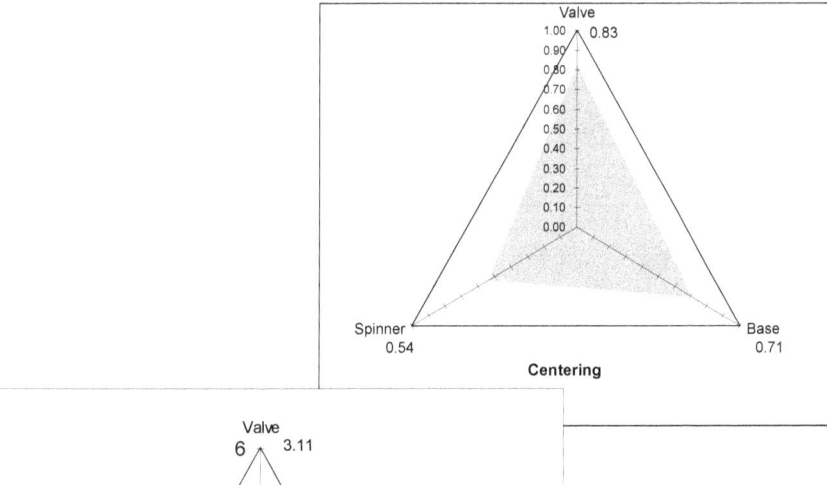

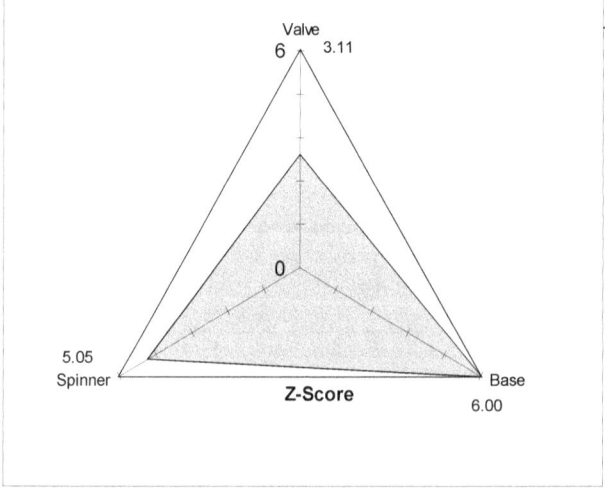

Source data for Z-Score graph are cells N2:N4 and Centering Graph from cells I2:I4.

.